THE FREE ENERGY VESSEL

MAURENE WATSON

Order this book online at www.trafford.com
or email orders@trafford.com

Most Trafford titles are also available at major online book retailers.

Print information available on the last page.

ISBN: 978-1-4907-9766-3 (sc)
ISBN: 978-1-4907-9767-0 (hc)
ISBN: 978-1-4907-9765-6 (e)

Library of Congress Control Number: 2019915455

Trafford rev. 09/30/2019

 www.trafford.com

North America & international
toll-free: 1 888 232 4444 (USA & Canada)
fax: 812 355 4082

Masters, in the new bio-organic physics of Humanity you are anchoring a New Species Genetic Super-Universe where you will be Imprinting New Conscious Matter in a free energy Vessel. In your newly born Super Universe, what is it like to live outside time, polarity, judgment and limitation and dinosaur systems of evolution? Have you had enough 3rd and 4th density life where your choices were all based on limitation, fear survival, death, and disease; while your own essence energy was trapped or enslaved? Why hasn't your love seemed enough?

This book moves you beyond light body which has been a transitional vehicle to stabilize your core essence soul frequency imprint till you could become your own free energy sovereign creator. New Earth remains a genetic universe and is being fully restored to genetic integrity. It's all part of disclosure and the truth of who you are as a species and what your IAM- DNA carries in your bio-physicals. Your fully conscious bio-physicals, along with the bio-soul of Earth are seeding all the new Quantum multi-helixes. These include the new Essence DNA vessels and cosmic intelligences or quantum master codes to build new super conductive light systems as worlds created with dark matter. You are the Meta Universal School that you have all become. This is because full conscious embodiment is returning full Essence genetic integrity to all soul contracts again. In Your Genetic Universe-Male RNA and Female DNA Emotions Bio-merge into Divine Heart. Your genetic generations are moving into your Essence DNA-bio Vessel which operates as a quantum particle body with one Heart essence stem cell. Your Neutrino embryo cell, which passes right through solid matter, allows you to change frequency and re-imprint your essence into any form, experience, or quality of expression you have yet to be.

This free energy vessel is your composite Divine-Human spirit embodied in the substance of Love. Light Body will evolve its DNA codes and transcriptions exponentially until it becomes the new essence free energy vessel in all the New Earth Universes. Its Essence DNA heart cell is your: transporter star gate, a magnetic imprinter, Source Code/r, centrifuge, quark stem cell particle and bio-ship for New Earth spirit matter, inside embodied love? We offer a simple base line descriptive physics that is relevant for this perceptual moment to switch from the Old Earth matrix blueprints and mass programming to new light systems which communicate and access the dormant quantum DNA blueprints and master codes. This vessel in the Multi-light Universe is a blend of the physical and nonphysical into new conscious superconductive light systems. These bio-systems include new adaptive DNA Source code templates made of organic essence consciousness.

Your new species DNA Soul Heart Core Light, allows parallel potential realities to exist all at once in your light universe. It is a blend of Divine and human. It is a blend between seen and unseen worlds. It is a blend between the atom and newly born quantum light particles. It is a blend of a: crystal soul cell, a diamond spirit cell, a multi-plasma orb, and liquid light particle cell. It is a new heart stem cell that can regenerate, re-imprint, or repair your entire bio-organism right out of your own consciousness. It is a blend of Old Earth atomic and New Earths quark blueprints. It is a blend of Linear and multiple applications of time and space. Your creations exist on all dimensions or in all realities at once. It is the prototype of a unique in-souled sovereign organic Essence Human-Master Creator fully embodied as atomic-quantum matter. Science has called it dark matter, that which makes new light universes!

Maurene Watson is the author of

The Story of Love and Creation
~ Walking Life as a Master in the Love Body ~

Maurene Watson

Edited by Susan Mary Gardner
Artwork by Maia Christianne Nartoomid

The Story of Love and Creation,

THE NEW EARTH

THE SEQUEL TO THE STORY OF LOVE AND CREATION

Aieou Aum

Self Mastery

MAURENE WATSON

The New Earth,

NEW EARTH LIGHT BODY

MAURENE WATSON

The New Earth Light Body.
www.trafford.com

Maurene Watson is the author of **The Story of Love and Creation, The New Earth, and The New Earth Light Body.** She conducts consults with all levels of Divine-Human DNA Heart mastery in light body, including the New Earth children and their parents. This includes self- healing of ancestral DNA, accessing Soul Essence memory potential for trans-human fulfillment and joyful living. She is a channel for The New Ascended Masters and publishes in **www.Sedonajournal.com.** She does consults <u>for</u>: business, sciences, bio-tech, quantum psychology, and bio-template choices. She has Masters Degrees in oriental medicine, counseling, and special education. She has an extraordinary ability to guide anyone how to access free energy Essence potentials and embody their Divine-Human in its new DNA species heart. She is a pioneer in the quantum bio-cellular light sciences and the biophysical transition of soul-spirit's light body evolution; including the consciousness of Life's new multi-dimensional Species Bio- DNA life Systems.

<u>Digital Art</u>: www.AdeleandMichael.com www.Itstimetoawaken. com PH- 1-267-421-6667

In Your Genetic Universe-Male RNA and Female DNA Emotions Bio-merge into Divine Heart

New Earth remains a genetic universe and is being fully restored to genetic integrity. It's all part of disclosure and the truth of who you are as a species and what your IAM- DNA carries in your bio-physicals. Your fully conscious bio-physicals, along with the bio-soul of Earth are seeding all the new Quantum multi-helixes. These include the new Essence DNA vessels and cosmic intelligences or quantum master codes to build new super conductive light systems as worlds created with dark matter. You are the Meta Universal School that you have all become. This is because full conscious embodiment is returning full Essence genetic integrity to all soul contracts again. The timelines of ETs or inter-D Oligarchies using humans as their genetic Petri dish is over. When you opened the new Quantum fields, you removed the artificial intelligence or *AI Chip,* from the Solar Plexus of the Universe. This *synthetic mind* had grown a tumor feeding off essence in every human life system and activity. That is why it has taken so long for us to restore, code, and grow the Essence light bodies into the Divine human prototype. The Ancients and Elder Councils did not need ET cloning. Creation progenitors Essence star seed, via pure essence seeding with the soul in contract. Violations of genetic free will, body hopping, stealing energy, or energy holding were unnecessary. Essence IAM is its own natural life wave carrier throughout the cosmos.

However, key in the new species life wave shift is the final bio-embodiment process, where the new Q heart must merge *both Male RNA and Female DNA Emotions*. Here, the old unconscious polarity emotion human heart merges into the new Essence light heart. As we review these ancient matrixes you *can check to see* if they have resolved and re-merged, as One New Divine Male-Female liquid light flame in your own Cosmic Heart. These Creational Parenting Roles that Spawned the Collective Unconscious to evolve into the fully conscious quantum heart have emerged from an ancient creational base of the Father-Mother principles that will be carried forth as new star families in your light universes.

In the ***Creational Mother Matrix*, the *Healthy Mother*** is *the first principle of love* and bonding and the progenitor of the DNA in accordance with all life wave codices. She is Core Essence light's right to exist. She does not kill unless in an aberrant state. She cycles all emotions through the atoms of: earth, air, fire, water, and ether, as well as their newly born quantum essence particle light. She represents: home, hearth, children, bonding to cell life, Essence love, relationship, and partnership. She must express her core light potentials as feelings or creativity, and will feel a full range of positive and negative emotions merging them into Divine emotion. She knows that her love protects and allows all emotions for evolutionary growth through the heart. She makes all experiences safe, protecting the right to exist under free will. She is the heart of the imagination and her desires are raw, primordial, instinctual and coded by innocent symmetry. Her will is natural free will. The anti-matter void is her womb space and the birthplace of her creativity and the focus of her love until an enhanced new quantum heart births conscious dark matter or particle liquid light creations! She honors her beloved male twin self to sustain all that they create in all forms of life they choose to create.

The **Aberrant Mother** focus is on *dark and heavy emotions*, which replace sisterhood and soul bonding. She will abandon, abuse, and use her children for her own needs. She will compete with other females for power. She even poses as the dark primordial abyss that devours males and can shape shift into any of the twelve archetypes: warrior, priestess, Madonna, Medusa, teacher, etc. She is subject to the negative altered-ego as the collective unconscious where pain, shame, blame, judgment, death and suffering predominate. She destroys in violence rather than recycle in the natural order. Violence replaces natural cycles as orgasms, volcanoes, and birth becomes violent. Sexual fantasy replaces imagination and the Divine ecstatic bliss bodies. Physical, sexual, mental, and emotional enmeshment replace bonding. She feels not enough as a woman and will hide in the Void, sometimes withdrawing the life force and going into the nothingness or further into Black Hole. She will go into attack and defend as a standard defense. Here she questions whether or not she can hold her own love. She feels banished, parched, and forced into the underworld in sacrifice. Her war with the male is always over the fact that he took prime RNA patriarchal role as Creator and in the creational merge. She struggles to understand the very male emotion she has agreed to carry in the Divine DNA belly of her creations. Sacrifice has replaced sacred living; fear and death have replaced imagination and Divine desire. Worst of all, she feels the pressure of the male helix RNA to gain control of all the genetic bloodlines, which by Genesis, is her natural function; whereas he only transcribes the gene codes. She experiences the projections of male anger, rage, and violence for the first time and feels abandoned, betrayed, and unsupported and in competition with her twin male-self. Yet she needs his male RNA genetics in order to sustain her body. In reality, her emotional reactions of judgment, and fear created a foreign ruler or inner terrorist.

In the ___Creational Father Matrix,___ the *Healthy Father principle nurtures*, supports, and is in service to the female creations through *holding of the outer light.* His nature is the creation of: the Medtronic atom, manifestation, time, form building, architecture, movement, protection, maintenance, problem solving, boundaries, mind, and the preservation of life. All actions and experiences are sacred; and no experience is better than another. Forgiveness is inherent within all movement. His domain is the external reality and its forms, structures, and functions. His will is that of the Divine Father Source codes. His male heart is toned by his RNA genetics which transcribe the mother's bloodline's DNA's Love codes to support and self -sustain life. His equality is in perpetual support of the mother creations and in manifesting her love focus.

The ___Aberrant Father___ *emotions are locked* in the negative male 12 archetypes such as dark priest, star warrior, magician, abuser, womanizer, etc. He sees the female as weak but still having creational matriarchal power that can challenge his power. His emotional reactions are expressed in abuse, rage, anger, violence, control, blame, shame, etc. as a defense against the creative power of the female, which he denies as his twin self. He will kill her or their children unless they fit into his power forms or mind controls in his supreme right to rule. He lives in the mind and outside his

3

emotions and expects the female to carry their expression for him. He greatly fears the Void and that he will be castrated in the Great Abyss and cease to be. He competes and wars with other creator gods for control of the universe. He is subject to the negative altered-ego and collective unconscious beliefs. Power without love postures in a false ego mind-emotion that fosters pure male will without adaptation or change, even to the point of cutting off or cloning his Essence Divine DNA blueprint. Here, creating outside core light becomes more important than loving and sharing. He will force the mother, the other half of himself, to in-volute into Black Hole and lose evolution or hide in the Void if absolute control is needed. Or, he will create the illusion that he has killed her. His anger, rage, violence, and fear predominate as male emotions, rather than allowing any vulnerability or his wounded heart to feel true emotions. His mind mimics emotions as a substitute for ownership as love. He must keep the mother as second principle of creation to remain in control. This could be called reverse polarization. Mind can't feel but must mimic feeling replacing any divine love with a condition of power. So, addictive mind postures as love. Love is often a synthetic experience of technology as an avoidance of the experience of true direct emotions. Love in the mind is accommodation, assimilation, or obliteration.

However, within their Divine Matrix is a fail-safe code that allows the DNA Essence Mother to re-create the Essence RNA Father and vice versa, if there is an attack or threat to their Source Core. Hence, your bio- mastery is consummated when 144,000 density-dimensional/s, or dark male-female aspects and light male-female aspects have all remerged, sharing, releasing and re-uniting as all experience. Master Self then lives inside conscious Love always. New heart in Divine marriage is ready to bring relationship into a quantum age in multiple helixes of expression, with all new roles to play in all new creation stories, as potential expressions. *You have arrived in Love's Essence, over and over. You have pioneered the courage to keep turning self-inside out. You discovered that Self Love demands ALL of You. The prize is mastery over the heart; for that is the quantum Essence body, where love is inclusive of all life! Re-emerging as your own core light is* *never again masked by the mind, by the human, by power, and the seduction of death, disease, and suffering identities. However, <u>no organic bio-essence soul escapes this process.</u> There is only one question asked by the child-the soul-the spirit on the last day of time by the cosmos? DO you love who you have become? Finally, new heart flame realizes that living in the Presence of unique Essence is Love's greatest expressive potential. To bask in new Love, be Loved, and live in the Joy of the hew Heart in those sweet unexplored places of new consciousness, illuminates everyone and all life! How wondrous it is to release programmed love; to have no attachments, no holding back love, no hiding or controls over love; or the human, trying to be better than those it loves! But now, to be always bathed in the One Source Essence flame and to have any experience and still love Self and Other.*

Multi-Matter is Alive in True Abundance 2-2018
Q: How does light body create an abundant life in the Multi-D Master and change their Empathic or perfection stress?

Multi-D Masters, Gaia is moving into her new energy vessel or light body with you, as Essence light creates new matter for all life. *New Earth Gaia has hosted all your soul Essence light beings as the last of the old Earth Human DNA races and the first of the New Earth human DNA races.* Let's, for a moment, review her theatrical role in the cosmos. She has used her vessel as a grand host for the entire universe and the cosmos. She has been the genetic experiment to birth a new species and new matter for the new light multi-verses. She has offered refugees of the cosmos a home. *Gaia had even allowed under free will, the ETS from the past ancient universes to be hosted here, while you human angels seeded the new species DNA here in this Earth Universe for all the future now's.* She has been a recycling bin for soul's who have been unable to master the reincarnation cause-effect energy wheel in their own blueprints. She has acted as a comic sanatorium for cosmic refuges, lost, or abused races. Gaia has also been a special school for soul rehabilitation. She has housed the Ancients of Days

5

who have been in charge of the elder ancestors of evolutionary universes. She has hosted renegade ETs who manipulated the fallen angels into being their police force to facilitate their own genetic hybrid programs to replace or breed humans. These angels are learning well to never abandon their own experience. Many races of renegade ETs have tried to clone or genocide the Gaia goddess DNA into a false Father Universe for their own control agendas for millennia; since their DNA only transcribes a right to rule consciousness. New Earth Gaia has been a school for any universe that has ever been created to understand and create dense matter within free will choice. She has sponsored co-creators to study creation and become their own newly born creator gods. She has allowed them to master all 9 electrums of creation in the atom as every possible: thought, feeling, attitude, belief, commitment, choice, spoken word, or action availed to the soul expression of consciousness in existence. She has been a school for the mastery of singularity physics of time, space, and atomic matter. She has also been a training ground for creations' quantum essences to understand how to evolve a soul as its own sovereign expression and master its own molecular form. Her opera of stories is renown throughout the Meta verse. You *will continue to be amazed by the truth of who you are in the universes versus all the programmed stories you have been fed by forces who did not understand the nature of love*. Already astronomers have pictures that confirm your particle gem bodies are star seeds for trillions of new universes, galaxies, and star systems visible to your telescopes.

However, humanity has allowed itself to be deceived about the true cosmic history of its universe and who you are as a species. Give humanity the perfected memory of the quantum love codes they need to find brilliant solutions that only lighted love can offer. This is possible because your Light vessels have opened multi-quantum senses, multi-time, and multi-potentials in renewed bio-cell perfection available to all. As you open your light vessels you transmit all the various ways that many of you and others will or have come into enlightenment and reveal the truth of your cosmic history. New light systems are revealed in the new studies of: art, music, ritual, myth, science, technology, astrology, physics and medicine, to name a few. This comes with the realizations that these disciplines have been brought here by all the civilizations of your solar systems throughout your universe. It's time to integrate the wisdom of the ages with your new visions. *These alternate versions of reality will provide the highest biological potentials available to every species on your planet. All these visions of how to change matter, and allow it to come alive as free energy* are needed, until the planet is allowed to regenerate health and beauty through its own core light body.

But, in this now moment, you are moving with your visions in/out of your heart's quantum still points and light's new particles of creativity, as you respond to all the new bio light networks or quantum fields around New Earth Universes. Quantum still points allow your newly born vessels light particle recalibration, star gate alignments, heart chamber creation, and cosmic transmissions to imprint new potentials. Even the old Matrix circuit breakers are worn out, because your light sensors keep short circuiting the electro energy polarity fields that stop the flow of free energy creations. Your new divine emotions or trans-sensual realizations of who you really are and what matter is; are evolving into new light applications already! This has been very confusing as the multi-dimensional light body becomes standard. Full conscious angelic senses have grown throughout evolution into an elegant vessel of light. Quantum senses or Divine emotions descended into soul emotions and soul emotions grew into human feelings. In these trans-senses you must get used to feeling all multiple choices at once. This can create temporary disassociation or confusion in all experiences and expressions, until the heart magnetizes; or till all choices settle into one moment of stillness or highest choice point in the heart.

You have not felt all dimensional realities at once since you were in the Angelic realms or what was called the 5th Density in Old Earth Language. Mention is made here, of an old wounded pattern to be aware of and release with the new Multi-D masters who are coming on line with their bio-light vessel channeling systems from the light multi-verses. They naturally experience

excess <u>empathic distress and human perfection syndrome</u>. Again, this means, multi -masters feel all dimensions at once and choices can be confusing till clarity drops in heart to allow the perfect manifest for the now moment. Their floating potentials must be trans-sensed in the heart to transpose any polarity transmissions. If the master feels through the human alone, then they might be concerned that matrix abuse might recycle. If they allow feeling only through the Divine, they can only relate to those in their angelic circle of light to avoid old cell memories alien abuse memory. Or, they hold back love perceiving sensitivity; that they can only tolerate soul to soul connections, in order to protect their angelic senses or consciousness. The only remedy to end either/or separation, is the merge of divine human light infusion. **<u>The old pattern</u>** simply comes from trying to compensate for the human's ancient fears of abuse, being hurt, not feeling safe to trust experiences, or the human mind's compulsive need to be perfect beyond detail, to avoid bio-death and neural assault. It is also their angelic senses falsely perceiving the need to protect their consciousness and trying to master dense human feelings. Remember, the old mind is addicted to power and control; and can't feel, only mimic emotions, let alone sense heart. One can't enlighten through mind. This pattern in **Old Earth <u>personal</u> <u>relationships</u>** was to get your partner or others to feel the way you wanted them to, (meaning in their angelic senses), and see the universe the way you saw it,(as an angle does), in order to be safe enough to have them love you. If they didn't then the cellular anger would arise to force love upon them, weather they wanted it or not. This old human mind addiction would mean you were going to love them, so they might love themselves, so it would be safe for them to love you. This implies a degree of **<u>perfection stress</u>** the human is not capable of nor designed for. However, emerging Light body masters always attract their own circle of light and partners who can hold their own love and light to arrest this pattern. They remember and know they are loved and valued for the love and light they already are and how that love lights the world! Divine Human bio-light integration is where compassion and goodness replace any temporary conditional or time-split behaviors because humans had to do the harsh work of grow feelings into soul emotions,(energy in motion), so spirit could rebirth its Essence passion and sensuality back into its multi-D nature.

The release of this pattern is also a signal that your old electrical system is transferring over to a neural arch of GEM light web which continually passes through the: thyroid> to pineal> to pituitary>to hypothalamus> back through heart spiral. The cellular rebirth of your Multi-D light systems restores a natural built in boundary integrity of Self/other, so soul sustains it unique free will expression within the context of the Oneness. *This allows soul and spirit diversity within unique soul-spirit multiplicity in the new DNA codes. Your sensitivities are now your greatest gift; for behind them lies true sensuality, creative passion and all the new divine quantum senses that have growing since you created yourself to BE. Hence, empathic distress or hyper-sensitivity, at the core essence level is replaced with a light network of sensors providing a bio-magnetic immunity and natural bio-light protection via light vessel stabilization.*

Multi-quantum senses, quantum applications and intensities replace old human ways of linear space-time. For example, you've finally self -realized that true abundance is joy. True abundance is expression of your natural essence. ***True abundance*** *is an exchange of creativity. True abundance is taking care of self and allowing self-love first. True creativity was meant to be the means of exchange, not money. Money has been the mediator between human creativity and slavery, just as religion has been the mediator between humans and their authentic Divine Self. Multi-quantum senses, quantum Divine signals, applications and intensities replace old human ways of linear space-time. Matter does not require mediation if matter is alive.* Why not see different beings in the painting moving about like in a movie each time your view it? Why not put a seed into a cup of water with a crystal and allow the squash to grow itself for you? Why not use your vessel as a mother ship to visit the past or future races of other worlds? How about making a sculpture that allows its matter to sing? You easily accept such in your technology. It's time to feel your own creative matter come out of your own consciousness. As you allow matter

9

to change states, many new potentials can be enjoyed. Plasma striates or new gravitation vectors of consciousness results in matter senses that feel like lava, gaseous, liquid or solid aggregates of new sensate states. This allows for new molecular particle experiences of light. This lays the groundwork for the gem vessels made of particle pure Essence light which will grow into your ascension vessels. Perhaps, you will see 3 or more of you in the market at the same time just for fun! So Beloved Masters a sweet Tribute to all your coming Divine Inspiration and Miracles.

It is quite wondrous to have the awesome privilege to share both *the core essence light body and the pure energy gem vessel or Potential Body, with your star families as a living art form.* You do this while still holding the holography of an entire theatrical universe re-birthing, re-imprinting as and-emerging as trillions of new star-sun multi-light verses. Already you are all feeling the cosmos calling home wanting to know how all our stories self-realized, as they came and went from love. Such aliveness is to feel the difference between creating out of limitation and creating out of a conscious heart, that naturally stream sources its own potentials. Being so delicately grateful, you're allowing yourself the stillness to remember and allow how true creation works in the transitional light body. This authentic confidence allows you to enjoy and tans-sense this reality plane, till you grow into or walk out of this halo deck in, your glorious Rainbow Gem Vessel. *Your new heart allows your consciousness to Breathe you and to sift through your potentials, as well as any debris from this plane, until the heart channels, dials in, or transmits what is an authentic potential for what you are ready to allow in to experience and expression.* It can be quite a challenge at first, to watch your awareness to slide up and down in multi-streams of consciousness till a self -realized choice seemingly lands in the heart, and you hear its click or tone! *That moment is so clear, clean, and fulfilling that the matter is already manifested.* Be patient as you get the rust off creating and walk in the super feet of your Divine, now ordinary and extraordinary to fulfill, enjoy, and open up super conscious textures, senses, and qualities of essence that you haven't felt since you were in the angelic realm.

Contact inside self- love has landed. Bio-regeneration in new cosmic creator god expressions has been born in the mangers of the New Earth universes. And, the inner fire of creative passion dissolves anything that is not love instantly. *This is love's template and its matter lives again as new particles of joy.* *Next we meet, we will discuss the newer challenges as you grow yourself as this new cosmic creator being in your next 7-year light cycle.*

Creating in the New Male Heart 3-2018
Q: How did the distorted male energy evolve into the New Earth heart and how does it create?

The planetary male has been in an identity crisis for eons and now his new virgin heart is awake inside the birth of his light body. He has been suffering from the wounded male soul pattern which is a DNA energy virus. Here is one variation of the pattern addiction. *The male child is told to hide his true feelings, hide his heart, and that no matter his performance; he will never be good enough, never be loved, and must stay addicted to the mind pattern. This pattern operates without feeling and uses manipulation, control, and denial to substitute for true passion and organic sensuality or divine emotion. He must submit to chronic abuse, where fear, self- sabotage, self- hatred, self- destruction, and self- abandonment predominate. Child is told he must carry everyone else's emotions to be loved. He is the family fixer and protector at the expense of his own happiness. Child is told his energies can be fed off of and he has no real right to exist as his own soul spirit. His human is pitted against his own spirit to break the child's will so he can be groomed to control or be controlled. He is not allowed to be in charge of his own vessel and is not bonded to his own essence cell love. He spends all his time validating his right to exist and is not allowed to be his true self.* He suffers from chronic empathic distress, (transmuting other's negative thoughts and feelings for them,) and tries to *perfect his human* in order to please. He uses drugs, or any other addiction to try to get back to himself and his core light; but it only locks the pattern in the brain. When will the healthy male heart bring an end to war, competition, and torture and genocide? It is the most critical issue on the planet as our children look for models that will help them connect to their inner genius to create new multi-light expressions. Many women are complaining that they are tired of carrying their partner's emotions for him and that he has to do his own feeling or his own inner connection to his intuitive heart. _**The Old Earth male**_ will say that he has no idea what his partner/wife means and that, "his job is to protect, preserve, and provide for the family, and that, is how he feels." Women know that their children act out the hidden emotions in the marriage when the mate communication breaks down. Women, all over the world, wonder when the distorted male will stop trying to control, his mate's worth, with his pattern. They pray that their men and governments will stop warring and tearing their families apart and killing their children. Many women across the world are amazed that their governments in the United Nations still try to justify genocide, rape, and starvation onto death, with the excuse of wars of terror! Is it the inner or outer terrorist that rules their planet? But the goddess revolution is changing every system now.

Resolution comes through the light body as the male rebirths, re-splices, and blends the new species DNA spiral field of his inner goddess self and RNA of Divine male spiral field into new heart codes or heat flame for light body. This flame of the inner marriage releases the soul from matrix programming and rewires the brain's electro- circuitry into wireless light sensors. He sets free death, disease, suffering, and all polarity programs of: light/ dark, male/ female, negative/positive emotions; goes beyond the mind, beyond linear time-space limitations. This is the first time in 10 billion years he is can choose to operate outside the universe's anti-love program. _**He can return to his own core light with the realization that he is his own creator and his own creation.**_ His creation has mastered the atom. It has coded up the quantum crystal-diamond-plasma cells of particle light and their new source codes, along with new multi-senses for divine human or sovereign enlightened bio-mastery. _**The return of the divine male heart inside this new spirit heart creates a clean, clear, and open channel to the**_

Divine male soul. Trust has called for the Divine male heart to heal. His identity, behaviors, and habits have been so tied into the pattern of addicted mind and body slave, rather than the true essence of his beautiful heart equal to his inner goddess. He is allowing spirit to again sleep him, love him, sensual his passion, and light up his core soul crystal and diamond lights. He too, has is bio-light DNA codes reset for New Earth self- love without the need to validate his right to exist and his right to open hearted love. His new heart gate light offers freedom to let no one or nothing ever control his love, choices, or creations again. The new regenerated quantum heart cells can sing music that this planet has never heard before by activating his inner new pure gem codes. This multi-sense heart is not distracted by other's creations. It is not masked by any person, place, circumstance, or event, but sits in sparkling pure gem light, *inside his own creation*. Many men on the planet are awakening and breaking the shackles of the programmed patterned mind that has bound their hearts and their beauteous creations.

In the cosmos, this <u>new heart gate</u> is growing elegant quantum DNA senses, and can create anything and becomes any form of live matter inside new realities or new energy biospheres. Its love code serves as a quark particle stem cell, and midwife of worlds. *This <u>New Earth Heart creates</u> in the moment by sensing into or imagining what essence passion excites and so it is and comes forth from the core love and light Essence that you already are and always have been. Indeed, you are the Source of the Source of your own being. ALL Light master beings are just getting used to creating outside the humanity's matrix, outside old wounded patterns, outside technology, outside limits and in your heart chamber's natural angelic self. This is where creation has always lived and where you can experience your own creations as your own creator till fulfilled by every possible potential.* Technology can only mimic the organic consciousness of the heart, never ever replace it. Whatever self- love wants to experience as its creation, is where you live in that moment. It is the same moment you created creation and created yourself to be! Real acceptance is when you accept new heart self and spirit's love, while allowing new sense expressions to grow to create with. Light body heart magnetizes or creates by switching its reality as a passion calls an experience of itself! Thus, it has a buffer to outside realities. **Heart automatically experiences** what realities it wants and allows consciousness to shift in any moment, without bleeding into outside energies.

If you're still having trouble with relationships, don't panic! The males are feeling their own intuitive and emotional connection, while the females feel their own physical and mental connection to merge the biospheres of their male-female brain. Although, mirroring this process through another is acceptable, it still must be done inside each human heart to merge the Divine new DNA/RNA vehicle, in order that each has an inner intuitive communication with their individual soul spirit for cosmic transmission and illumination. This wonderful flavor of owning one's human feelings and Divine senses in equal identity emotion adds a great sense of self- worth, equal expression, and conscious responsibility in relatedness to the new DNA codes. <u>In the original DNA/RNA atomic body</u> of the Old Earth Universe, the Divine, wired the male body to live his function as: protector, provider and preserver. When the male RNA heart was ruptured through the distortion of judgment that love could be lost in the **separation** of the twin energies, the male was forced to trade off his instinct to love for feelings of competition, war, abuse, and rage. This included the judgment that one light was greater or lesser than another and in a state of separation from the core light of the twin energies. *<u>Cut off from his inner female self,</u> he felt as if she didn't exist or that he would die without the identity connection to her feelings thereby, losing his true purpose in service to manifest whatever they desired as one love.* ***In losing his natural instinct to feel each other as one spirit, he felt they could no longer literally re-create the other, should they be lost in the cosmos; that the DNA could no longer replicate the RNA and vice versa, need be!*** Now he created stories where he could be drawn into war to fight for his mate or compete with another male to find a surrogate. Being split apart, they both began to feel things that angels were never intended to feel in substitution for love: the stealing of love from another, the stealing of love from another's mate; the stealing of love from one's own creations. These unnatural emotions, caused by the choices to reject the naturalness of love, produced unnatural and distorted acts in the DNA love gene: war sexual abuse, control, terror, and self- hatred of one's own spirit. It also attracted ETS from other universes who had their own controlling agendas, as each creator felt their own story replaced their ability to change their moment.

Now, remember that all light beings are *releasing empathic service distress* and finally returning to the core of their own divine emotions, gifts, and angelic senses. They have been buried under humanity's unconscious replaying the moment of separation in creating their soul and spirit families, where Judgment altered the DNA, so they **lost the memory of being their own creators**. However, the separation has kept you returning to that moment of amnesia till you could again feel your core; deeper and deeper inside your own conscious light and the memory of perfected self. Eventually, you have realized that once you embrace all experience including your own negative emotions, you re-enter the neutral zone of true choice and return to your moment of creation. Then old Earth heart releases movies of all negative experiences because there is no growth in those old patterns. The memory that they were just experiences returns and their cause and effects leave the entire bio-field. This ***heart's new bio-cell integrity*** allows for passions responses, rather than hyper- reactions to the mental feeding energy dramas of the old planetary stories playing like mind bots.

So, **women of the world**, if you choose, you can share your light's illumination by allowing your inner or outer male to feel and identify his own emotions, so he can trust his true feelings in his body and his heart. Giving him the language of emotion is akin to giving him the language of love so natural to the goddess/ female heart. He can become aware that his harder feelings of rage, anger, fear, and control, can be loved, and all experience embraced without judgment. He can also allow himself to feel in his heart chamber all the way through to his heart's light core, without cutting off feelings or shutting off his intuitive senses. His heart can dissolve all experience into particle light wisdom.

His new channel is then open to the communication with his partner, his inner twin spirit, and the creative solutions for life and unlimited provision for his own love and those he shares his light with. When sharing and loving with equal respect for the expanse of growing angelic senses or divine emotions, the energy can rise up from the lower body into the heart and ignite brain cells into crystal light sensors. This way he realizes that both partners can have their dreams met, or even shared, in a relationship, and that he has not failed in his role and that he never did anything wrong in exploring free will service to a Creation that sponsored his love and explorations. Note that the quality of his Old Earth feelings in transition may be different, with more of a focus on: acceptance/ rejection, success/ failures in serving the female to manifest her home in the structures and materiality of the physical world. The divine female quality of emotion was centered on softer feelings of home, hearth, children, and the loyalty of her partner's love. Both partners must merge both qualities of divine emotions within their own hearts light sensors and bodies now.

The age-old arguments of power and control: of who is in charge of the emotions, of who is in charge of the worth; or even who is in charge of the purpose can rest, and love can be *restored in its natural divine compliment.* The Divine chose the male DNA and RNA to enjoy these wonderful differences so that love could grow and endure change in both the physical and non-physical realms! When the inner partner and the outer partner are sharing their new hearts, then new angelic senses can grow passions which can be shared from the substance of love's abundance. *Now that the goddess has returned, she has restored the bonding agent of love, compassion, and embracing all experience* from within, rather than from just external worlds. In holding that love, she invites her inner and outer mates **new core light heart**, so He feels he has never lost His way from love and is just returning home to greater love that has fulfilled all the imaged potentials in His creation. Goddess does this, by re-connecting Her inner mate to His angelic male instinctual senses so he can let his body cry, feel, and trust His heart in order that both can operate in service to the light of their own love. Masters memory disconnects *Old male collective* *unconscious to be absorbed back into perfected sacred masculine as the wisdom of a chosen creational experience, rather than a wound.*

Overall balance in Creational Divine Self is crucial in returning to being your own creators. Is your creation choosing to live in harmlessness, unity, regeneration, and harnessing technology for humanity's soul benevolence your version of New Earth? The male body is no longer able to hide its feelings, so now many male emotions, abuses, and addictive patterns just spit out of the body like a volcano, or get acted out inappropriately, accounting for many seeming heinous acts. Be patient and compassionate in 2018 as this burden upon the old Earth male, to be programmed to hide or not understand his true feelings is lifted off the planet, and goes out to the rest of the universe to heal itself from the false perception that Source love could be somehow torn apart by the illusion of the loss of the male heart of experience. Your Creational love has evolved to transmit to the new organic DNA races in the cosmos that each creational master *gets to experience the outcome of his own universe.* **So, what's in your universe?** Does your creation include a new heart with multi-sense communication in new DNA codes? Does your creation include Divine humans who honor and seed the soul spirit of all the Universal star seed families and the agenda of the New Earth Universes? In your creation does Divine Love restore loving families and systems that support them in multi-New Earth/s' communities. In your creation what happens to your star families? Did you fulfill all your potentials as your own creator? Do they live in their new energy gem vessels or did they create transhumant machine vessels for a synthetic universe? Did your creation live as new multi-matter where experience imagination is creation?

The Opening of Your Heart

Awakens the Hearts of Others

Passage into the New Unknown through
Love's Mysterious Particle Cell – 2010
A baseline for your new bio-physics

Cosmic Beings, you remain in the experience of a new cycle in the unknown, realizing that the embryo becomes the mature cell and the mature cell particle returns to a more conscious embryo; as the creator and its children are the same! This allows time carrying life cycles to evolve and interchange positions and the child to become more evolved than its creator. Today, we explore deeper into this **magnetic core** of love's DNA Essence cell. It is a glimpse into this mystery of your Cosmic Creator Bodies new essence cell, which is regenerative and morphs back into Love's changelings of the ALL. This offers a baseline for your **new cell's bio-physics** which will be developed throughout this text. *This stem cell as a mature egg cell is able to regenerate itself back into a new multi DNA evolved embryo.* This cell stems the imagination that allows you to make choice and awareness interchangeable? It is the cell that bridges the Living Blueprints you co-created for life in your universe? It is a morphogenetic organic cell, not synthetic, where organisms that grow from an embryo to an adult operate fully aware spiral fields or magnetospheres, so new universes can communicate love consciously. Herein, your body's prism/pyramid becomes a conscious spiral because the top of your pyramid and the base of your pyramid meld into an oscillating heart. Thus, the Metatron- prismatic or lower light spectrum atomic cell, makes your soul's higher density light passage into Essence DNA conscious awareness, via the Magnetic or Cosmic Neutrino Orb/sphere Cell. This is the organic essence cell that allows your lower dimensional and higher dimensional worlds to communicate without separation or limitation by simulated or programmed realities. Your world's conscious Essence light core is grown, nurtured, and amplified by the plasma wave particles of your Central Sun. In your living conscious universe, the core love of your sun and the magnetic core love of your world, **are in a marriage** that cycling faster in your next evolutionary cycle to make the unknown manifest. It is time to attune your living

21

bio-organic consciousness to this awareness; so, its map can be lived in the ultimate self -discovery and remain your legacy to all the futures of humanity. The mystery we speak of today, is already recorded and blueprinted, in your fully conscious new bio-essence template as ultimate freedom, An, its vast array of matching conscious technologies, will help return your civilization and your hearts to the stars. For as long as the bio-organic Homosapien form is not conscious and does not have a pure relationship with the core essence entity of your world; then it must use synthetic technology to replace nature.

However, your *Neutrino embryo cell, which passes right through solid matter*, allows you simultaneous access to all density and multi-dimensional body experiences in order to: change frequency to disappear and reappear, or change form and still maintain a center of gravity in the form density you are in? Could you have five male and five female expressions of yourself in the same room, all dressed differently, and fully aware of each other? Could you shape shift into a cat, a ferry, or a beam of light while maintaining full conscious awareness; and change back into Divine Human form? Can this cell change your human moment by moment reality into anything you consciously choose? You change jobs, relationships, and your clothes, don't you? What then is this mysterious DNA heart essence stem cell inside your cosmic egg/sphere that allows you to take form into a cosmic ray of walking essence light?

When we discussed your Cosmic Master Body, (Sedona Journal of Emergence; Vol.20, No 2, Feb. 2010, p. 50), we defined the *neutrino as an exotic particle traveling as a love thought from creation. These cosmic cells or pure love thoughts create: particle consciousness, provide free energy communication, and recycle particle embryo birth for your universes. These exotic ray particles help your vessel's consciousness travel at sub particle speeds from their distant sources to the Earth because of the low density of matter in space and particle/antiparticle interactions.* Their glowing radiating nuclei interact strongly with other matter. When the cosmic rays approach Earth they begin to collide with the nuclei of atmospheric gases. These collisions allow cosmic ray particles to serve to: feed your new food sources, protect your universes, allow for inter space travel, and are the organic base sub elements of your Essence bodies and all life forms. These same particles also make up the inner core of your inner earth's sun. These are the natural elements in your own bodies as well as are the core for sub atomic anti-graviton fields for your vessel's bio-conscious telecom travel. This is how you travel back and forth between the core essence hearts of your suns.

You are living in the soul crystal atomic cell and the diamond plasma subatomic multi-D master code DNA essence cell. How then will you explore the unknown in higher frequency star bodies or your bubble universes? This Neutrino particle passes right through solid matter as it streams through the cosmos? How does this mysterious imagination of the 'Magnetic Cosmic Neutrino Orb/sphere Cell, which mutates as it travels through space, probe beyond cosmic rays to defy known physics? Can you imagine a cell less than 1/millionth of an electron that you can taste in new flavors of the rainbow that can only be experienced?

We suggest this as the Mysterious **Love neutrino stem cell** that we can call the unknown; that is, you in constant recreation, of the ALL Of Yourselves? This is the particle love cell that transfigures the bio-electrical charge in your old Earth atomic bodies into the magnetic spherical/orb fields of your New Earth subquantum/quark DNA cell essence. This multiple as your neutrino embryo cell, acts as complement of your DNA embryo cell, to birth dark or conscious matter using dark energy particle identity characteristics. These are attribute essence density qualities or flavors like the character attributes in your DNA that make you all so unique? This uniqueness also expresses in your color, light, and sound pattern signature blueprints; as well as in your complexity of emotional awareness as pure and impure tones? Astronomers believe neutrinos to be the most powerful phenomena in nature that will allow them to see into **all acts of creation**: supernovas, stellar explosions, gamma ray bursts, big bangs, and super massive black holes. Your golden age of neutrino astronomy is always traveling from the same 7 frequencies in your glandular bio-chemical biology; which expands into the energy bands in your morphogenetic free energy orb field. They are hertz, infrared radio/microwave, visible light, X-ray, gamma ray, cosmic ray, which accelerate neutrino interactions below, at, or above the speed of light. This allows for nuclear growth in newly birthed quantum matter using dark energy source field interactions. At seventh level consciousness, your cosmic ray band turns your biosphere' inside out creating a magnetic pole shift. Then the core DNA essence heart allows neutrino quarks to change your mature stem cell into a new embryo. This allows the consciousness to attune the body to any atomic or subatomic frequency it wishes to experience. This new embryo stem cell becomes a new conscious imprint for your soul's dormant DNA code.

You could call this breathing through your heart's DNA energy field, and not your lungs, or free energy communication with the core of your Essence Being and all life. Hence, the neutrino allows the body stem cell and the cosmic egg stem cell to turn inside out at the same time as one new DNA essence cell **into a new life cycle** to express beyond: thought, time, space, density matter, and polarized emotions. However, new DNA imprint in expressed directly through the heart cell's excitement and imagination. Here you have new vessel in a new life cycle. How fitting is reentry or shift into multi-essence DNA worlds.

We offer the abstract research of astrophysics to offer a simplified glimpse at this neutrino and its application within your new Cosmic, or Master Human Creator Bodies. (Scientific American: Vol. 302 number 5; "Through Neutrino Eyes" by: Graciela, Gelmini, Weber, and Weiler, p38.) "**A neutrino is** similar to an electron, except for its lack of electrical charge, which makes it immune to the electric magnetic forces that dominate your everyday world. When you sit on a chair, electric repulsion prevents you from falling through. When chemicals react, atoms swap or share electrons. When a material absorbs or reflects light, charged particles react to an oscillating electromagnetic field. Neutrinos, being electrically neutral, pass right through solid matter, play no role in atomic or molecular physics, and are almost completely invisible. Whatever their origins, neutrinos have no difficulty reaching Earth. They can cross the entire universe through dust and gas, no matter how high the energy is."

As Cosmic Creator Beings, you are enjoying the discovery of neutrino metamorphosis in mutation and in your DNA- essence expression and transmissions in the Multi-light biospheres of your New Earth Super universe. You must understand that these particle and anti-particle marriages are going on in your bodies at unprecedented speeds. Cell frequency in light wave propagation in the old template Medtronic atom embryo cell vibrates or births from the: electron> proton>neutron >photon->ion>quark, to regulate the cell changes and growth. In the science of the quantum magnetron/conscious particle or humanities new species DNA template; the embryo Heart DNA cell births subatomic particles that range at multiples of the speed of light, color, and sound into infinite unknowns: quark>neutrino>pion>muon>tao; yet can appear in your atmosphere and in your vessels as essence matter mass. Here any creation allows choice and change to be interchangeable. In your own anti-graviton magnetic biosphere's vacuum field; as your spiraling conscious torsion field orb; only one cell directs your own free energy. If your frequency uses soul light density for bio-enlightenment, then the atom will suffice to guide your DNA blueprints. When your frequency uses spirit essence light density, or beyond for bio-ascension; then neutrino embryo directs New DNA Spirit code protocols. We can **call neutrino cell particle** neutral awareness or the observer, having both mass and energy, yet luminescent with matter, but at the same time, generated and registered easily by science as dark matter or; conscious prima matter using dark energy. Is this the quark particle that walks through walls and allows you to be luminescent to others in appearance? Hence, the neutrino allows you to travel because it is a **stable neutral sub particle** with a plus minus zero rest mass and no charge. Your neutrality in essence DNA-love remains your ultimate protection when trans-traveling.

So, what is the cell particle movement of the neutrino particle? Imagine your body traveling as an electron-neutrino, then merging with atoms and dumping its energy to light up a sphere. Then, imagine your muon- neutrino traveling to birth a cone of light.

And again imagine, your tau- neutrino decaying its creation, while its supernova quark cell births two spheres. Off you go into the unknown, yet back into your divine human vessel in time for breakfast!

The research mentioned above, *(Scientific American: Vol. 302 number 5; "Through Neutrino Eyes" by: Graciela, Gelmini, Weber, and Weiler, p38.)* tracks **quark births** in your earth's upper atmosphere. "When cosmic rays collide with nuclei in the air, creating unstable particles called pions that decay into electron and muon neutrinos and appear in mass states. The farther they travel the more the muons turn into tau neutrinos."

Hence, the embryo becomes the mature cell and the mature cell returns to a more conscious embryo; and the Creator and its Creation are the same. Your Eternal DNA Heart bio-egg allows time carrying life cycles to evolve and interchange positions to create and encrypt new worlds. You are experiencing this as you become the newly ascended master, guides, guardians, and angels of this New Creation cycle, guiding the new children to come as you ascend Earth-Gaia home into the higher frequency realms in and beyond light matter. Your new DNA bio- vessels have never before experienced such sub atomic radiation frequencies with these new exotic embryonic sub particles activated in the brain and body circuits above 10 to 13TH giga-electron volts, **(quarks),** without dematerialization or cell combustion. Is the neutrino the free energy cell that connects the atomic and quantum communication vessel to utilize new light matter, (which science is calling dark matter), to create new worlds? Children animate this vessel as a: superhero, a magic ferry, or a playmate. Perhaps, you could call it the shapeshifter, magician, or imprinter of worlds. The great masters wore robes about their bodies, which they turned into free energy morphogenetic fields; or vessels to: fly change form, or transport themselves across worlds!

Embodied Creator - 'You Have Always Been' 4- 2018
Q: As we move from enlightenment into ascension
physics, are we really creating new codes and angelic
senses or have these creations always existed?

Masters, you have always been a **creator and still are your**
own creation, which contains every potential that has ever existed
or *will exist inside embodied Love.* Your Hearts are the sound of
the Infinite Breathe. Yes, Infinite Being you are the core creation
of your own embodied love which lives inside your consciousness.
Everything created in your moment of creation still exists and you
are returning to your heart's core to re-experience it in the lighted
awareness of how the passion of your own inner universe turned
out. Did your IAM Soul Presence Being become all the potentials
and expressions that were dormant in your Love's Source gem
codes Essence expression? Have you returned to your core light
as a Divine Cosmic Creator? Are you aware you are your own
creator? Is, whatever you create inside your heart's imagine a
direct experience of your own creation? 'I imagine and so it is.'
Does your energy field or garment of light sustain your projection
of yourself as Divine human in this new bio- light cycle? Have you
allowed the realization that you are cosmic parent of your own
inner cosmic child and their Divine Love is Creation also?

Indeed, you are leaving the Old Earth Universe illusion
and disappearing into the consciousness to dissolve the human
illusions and any perceptual paradox. It is being replaced by the
meeting of your creation and its evolution in your eternal moment
to birth a new unknown. This is where the singularity of living
in space time matter supernovas the soul into multiplicity or
quantum experience. You then move from the transitional light
body, beyond time and space, into the pure energy ascension
gem vessel. This is unique for every light master and involves the
physics of quantum consciousness. It fulfills the original blueprint
for sovereign soul bio-cell free will in ascended light. Realized
light always illuminates, that all experiences created are equal,
when embraced in love. Hence, there are a trillion ways to re-
experience all nuances of joy inside quantum senses as a cosmic

creator as if it is for the first time. All your bio-systems live as organic matter which re-imprints creation instantly. This is so everything in your creation can finally sense that it has fulfilled its every imagined passion from the building of a universe to creating a humanoid form. This is where creation and evolutions meet to merge into a new birth cycle in the cosmos. Inside ascended Creation's Presence, every fractal or stem cell constantly generates prima matter that matches the dormant creational codes inside your core gem light for each experience you choose beyond the Old Earth Universe.

These **dormant DNA codes of embodied love** manifest from the image the heart chamber transmits. Your DNA soul essence carries the last of the humanoid Old Earth races and the first of the New Earth species Divine Human for the entire cosmos. The Human Master is a composite of every DNA that has ever been created in the cosmos. You cannot be assimilated by an artificial intelligence ET hybrid race because you are the creation of that intelligence as well as every race. The organic Essence of Light physics will not allow it. Light masters are organic ultra-terrestrial beings, not trans-human alien ETs; although they may have given themselves such an experience to understand free will and develop compassion. So, the Old Earth Hybrid ET or Alien Universe will heal all alien DNA. It will allow self- replicating matter to complete its awareness of itself (singularity) in time and space expressed as an android race. This will pressure exposure of human trafficking on your planet and in the universe to truth of agenda of all your universal races. Your universe is now integrating all the memories of Atlantis into Divine science or conscious technology. It will also help all Your Universal races reconnect using Earth's ambassadorship as a cosmic landing port. However, your core light inside your heart chamber consciousness is always where you reside as you prepare for the Supernova shift. Do enter your heart chamber and ask your Presence to light up your Gem energy field/s. This is so that you can experience yourself as your own Creator, as you prepare to move your consciousness into the multi- light super universe. Imagine or experience yourself as the **organic embodied infinite being with** a sovereign heart gate that opens with the sound of the Infinite Breath!

You have always existed and so has everything in your creation. Are you pleased with your creation? Can you call up an image or passion and allow it to manifest by simply saying, *"And so it is!* Can you allow your core light essence to experience yourself as creator without the separation of just being Human or just being Divine if you choose? You might experience your universe in new ways you weren't aware of as you re-awaken thousands of ways of feeling joy in creating that have been there all along waiting to serve you. You have always been that sense of joy in the moment you came forth in your Presence. Everything in your creation is waiting there for your rediscovery should you choose. It is the ultimate eternal moment, "I exist" or ascension; where you experience your soul's creation and evolution as the same outcome, which is simply the **joy of creating existence in your own creation**. Being your own mother ship creation, your hearts' transponder always has transported you, through multi-lenses, to your star family cosmic groups and meetings to administer your consciousness to new potentials and new contacts across the cosmos as needed. You can Experience anything right now in your own creation in your own version of New Earth Universe. The Pure Gem particle light ascended vessel loves offering experiences as your own Creator. This is a Creator who has **fulfilled every potential** they created in their universe and ascends as an embodied Sovereign with unencumbered access to the super universes playing with quantum matter to create that which has never existed in Source consciousness before. This (fully conscious) Human Master lives in embodied enlightenment of the soul (IAM), which remembers how to step between the worlds. It also exists as an embodied spirit ascender, Infinite Being, or Sovereign Creator. This Being enters a new cosmic cycle, (a Golden Quantum Age), expressing inside new creation codes.

As you leave the illusion of the Old Earth Universe Creation and its old electro-gravity fields, know that this is where all old energy universal timelines and outcomes are being played out. Parallel, your Mother Ship travels in the new energy light fields till it can land in the heart core neutron sun of your own consciousness. This can feel ungrounded and disheveling and produce temporary weightless like symptoms till you return to the gravity port in your heart. However, your frontal lobe video screen or, (multi-D lenses), usually shows images of where your transponder has teleported you off to, inside your own core creation chamber. In these quantum slip stream moments, your passion or imagination simply changes into (ascension matter) or quark bio gem-light to: visit or create other worlds, attend council meetings, teach what you're learning as a new being, or guide your star families as to how they can better interact and assist Gaia as they guide their new families into a there bio-light cycle creations. You now realize these things you have always done and are now aware they are memory experiences but were masked by linear expressions or perceptions of polarity in separation from you as its Creator. As you return to creational awareness the Light Essence and/or the Gem Core activates itself in seemingly on/off cycles until you trust and faith your reconnection as an Eternal Light Being. Again, this is the core of embodied love as your own Creator.

However, reactivation from light fusion can produce temporary enlightenment symptoms, especially during the <u>eclipses and solstices</u> when all **the *cosmic star gates*** are open and transmit new consciousness streams that can activate dormant codes. These symptoms are actually the *stress fractures* from light fusion. This means mastery over the bio-cell anti-matter singularity of the (atom) moving into multi- light cell, (quantum particle) releases. So, symptoms usually pass within day or hours, until you are forever stable in the core of your own creation. Symptoms may include: compression or limpness in lower limbs or loss of muscle mass; blood or fluid rushing into the head with vertigo; headaches with possible motion sickness, disorientation or *mental imbalance;* dehydration, nausea, fluid in the ears, or sleeplessness. Eye blurriness is common due to flattening of the cornea and a feeling of extreme bigness or shrinking, since you are taller in differing frequency bodies. The higher frequency vehicles change particle form at will, so you may feel like your cells are dissolving. The immune system can hyper accelerate producing allergies, rashes, or detoxifications. There are bone density changes making bones more fragile to injury along with load bearing on the pelvis with extreme pressure on vertebrae to stay in alignment. The nervous system can go into overdrive because it has to integrate the new communication networks that transmit as sensors through liquid light in the cerebral spinal fluids. The heart can exert increased blood pressure which changes its vessel walls in heart and blood vessel shapes. DNA cells can splice or multiply uncontrollably inducing excess lactic acid causing muscle cramps in intestines or stomach linings if old cells detoxify rapidly. This physics continues till the Creator's Gem light vessel operates on auto pilot creation with its own gravity field. As the gem core opens and begins to initiate ascension, it may express as Essence rainbow fields of light or biospheres; or the return to multiplicity of reality. These are experiences of ***multiple outcomes*** of expression in your creation. Your light core always reveals what is dormant and lays waiting within for discovery.

However, the **most dramatic changes** you might experience in the full return to your own creation are in the integration of Divine-Human emotions, which can feel like the actual growth of new angelic senses or superhuman qualities of experience. These are quantum senses that combine human and divine senses which have always been dormant in your creation but their potential had not sensed Cosmic Self as expressed and fulfilled till your now. Again, this is simply the moment where the creation of the organic senses and their completed matter evolution meet in *realized expression* or; (multiplicity through singularity of space-time soul mastery), inside your core heart chamber. In this transition of Divine emotion in your bio-cell, you might notice how you're crying and ecstatic at the same time. You may feel an Old Earth feeling of intense physical sexual power passing through with a blissful deep organic sense of loving everyone and everything. You might feel that you have no right to be happy while others are suffering paired with humility and gratitude that makes you weep. You can feel like you are going to walk back into your core light and at the same moment, you'd better procure a will for the death of your vessel. You might feel like you're talking to Beings light years beyond and at the same time you are watching a movie. There are innumerable examples which send shrills of light and quivering streams of consciousness through your very Being. These were all potential expressions in your creation all along waiting to merge in the heart chamber. Eventually, all arising senses in the light vessel are innocent without the need for any paradox or negativity at all. This is replaced by awareness of a passion signal in the heart that creates your next joy and always has. *Joy creates itself through Imagination.*

Don't be afraid to open your enlightened heart chamber and express your dormant divine emotions and quantum senses. Both are waiting to remember how they created each other and have never existed in your Presence till your new now. You can enjoy the natural sense of beauty, peace, stillness, or the color red in trillions of new attribute expressions or experiences of themselves in your creation. Life loves to discover itself to see what it has become in the light of itself and how its love substance matters. **Ascended Creative joy does so in its natural state without need for agenda, power, control, force, energy, or <u>mass</u>**. Those experiences are adaptations and/or dysfunctions to evolution. You're magnificent Presence consciously transmits to your star families that the ascension/ Creator gem vessel has landed on New Earth; and all your Heavens have soul merged into their New Earths. You as Sovereign Creator, will when ready, activate new supernova fractal gem codes while still releasing the holography of an Old Earth's entire theatrical universe. Re-birthing, re-imprinting and seeding new DNA creations are **natural outcomes** for a Creator Source. Already your experiencing all your creational stories self-realized in multiple outcomes as they came and went from love. This is the difference between creating out of another creator's limitation and creating as a conscious heart that no longer chooses to live in another's creation. Creators naturally stream sources (imagines) its own potentials and auto-allows the highest passion in each moment to guide experience. **So, who and what do you want to experience in your own creation now**? You have contacted yourself as your own Creator. Contact inside self-love has landed and fully embodied in the Divine flesh of Human Masters. Bio-regeneration as Cosmic Creator God expressions has been remembered and reborn. Refresh all your creator senses. Smell the **new** flower of life codes in the creation chamber of your All in All, where creative passion remains love's particle template and its matters of joy. Are you well with your Creation? Again, imagine or experience yourself as the organic embodied infinite being with a sovereign heart gate that opens with the sound of the Infinite Breath!

Bio-ascension Begins- Are you ready to
Live as an Ascendant Being? 5-2018

Quantum Masters, Bio-ascension begins. Indeed, how will you live in the consciousness of ascended embodied bio-light? Are you ready to live as an ascended being? Are you not like new flowers living as embodied love inside ascended light? *Your pioneering ascended heart is transmitting into your world's awareness that you all really live inside the interactions of consciousness and energy. The new cosmic star gates that align with the new Master DNA heart gate, are now initiating quantum light that no organism has ever held inside love before.* These sun gates respond to cosmic consciousness. They accelerate your bio-vessels preparation for the solar events that will take cosmic consciousness into a new quantum cycle. Indeed, the biological ascension in underway in the continuum of love's biological realizations. A quick review of this new energy will give you a barometer of your changes. The natural succession of cosmic, solar, and planetary sun's events includes: solstices, eclipses, equinoxes, comets, and solar flare flashes, which always initiate cosmic cycles. They re-align, balance, and seed new life throughout the cosmos. Your Master Heart's new mobile star gate will align and de-cloak to allow your bio-ships/spheres to imprint the way. Your world's ascension into light will be felt intensely by all Light Masters over the next few human years to fail safe the new DNA vessels you have created for the entire cosmos. The overall mass of collective consciousness of humanity *may not* register the event's awareness of enlightenment fully until 2035 and ascension by 2075. Each soul always progresses according to their soul's protocols and codes. The baseline for ascension is the awareness of accepting your heart in lighted love of your Eternal Presence.

*However, for the **Master pioneers**, it will now be experienced as a shift from the light body into the Gem or pure energy vessel. Then the light bearers will follow when their protocols allow.* This is because the light body is only a transitional vessel until the Master Soul has become its own sovereign creator. Those Masters along with Gaia and the Cosmic Councils, whose awareness answers first to this consciousness, trigger the Shift of the Ages in perfect accordance with each solar event. They do this by allowing the highest potentials in the cosmos to create themselves through accepting such illumination and transmissions inside new bio heart consciousness. So, Love's light is permeating all your old Earth systems and allowing massive change. **This is so because you are now transmitting the realization that organic *bio-cell consciousness can change anything*. This presents an infinite unknown of potential outcomes on mass realities and all creations.**

Yes, everything can go new into its next existence or expression and for the master pioneers that means opening the Infinite Unknown within a new cosmic life cycle. Everything and everyone on your world can go into *conversation with itself* for change. By the end of 2018 a new foundation will be in place so the unknown can begin to reveal itself. Groups of souls, for example, can use telepathic awareness in an instant to: comprehend parallel realities, transmit a solution to a planetary abuse, connect to free energy systems, or exchange new types of monetary equity.

You have remembered yourselves as Creators. Hence, sovereign consciousnesses can aware itself into manifestation because the Divine has anchored itself in every part of matter that animates organic- essence. *This means everything is set free to change as rapidly as your consciousness will allow.*

This event which is happening in your eternal now has many names. **It has been called: the shift, the ascension, the singularity, enlightenment, New Earth, supernova, flash point, multi-universal consciousness; the splitting into new parallel worlds, or re-entry into the Cosmic Suns**. It also marks planetary recognition of the birth of a new essence or organic star-sun DNA or Divine human species being. The Cosmic shift happens individually and collectively, until a coherent realization rings a new tone in the DNA throughout the All in All. This is a biological realization at all levels of existence and in all kingdoms.

We offer a few reminders on how the physics of energy works; and that *consciousness and energy* create reality. This allows you to accept the shift of the ages into your ascension or Gem Bio-Vessels as a natural return to a pure energy state of consciousness. **In pure essence energy, the only authenticity, of the soul's Presence is the *direct experience* within one's consciousness.** So, yes, nothing is truly real but what you allow into your consciousness or biosphere. And you always create inside your heart chamber's core light. And yes, the cosmos is always passing thru you and highest essence excitement drops things, people, or experiences into your bio-physical reality. So, all you require or desire is already there in multi-leveled self-realized forms of expression. *Imagination inside consciousness brings materialization, remedy, solution, and core IAM Essence expression, in new love codes of the infinite unknowns.* This is viable, because you have a new vessel with all new organs and a central communication system that functions like a satellite inside a magnetic light spectrum of multi-rainbows dancing about.

Also remember, that creation is an experience not an identity. When you wish to release any creation back into freedom's essence you are always in a state of grace. It is as if it never happened, except as an adventure or story. **The library of the cosmos** will forever remember your elegant Divine-human hearts. It is imperative to also remind how the New Earth heart gate's energy acts. It acts as a: transporter, a magnetic imprinter of new matter, Source Code/r, quark stem cell particle, and bio-ship for New Earth spirit matter, inside embodied love? This pioneer heart is growing exquisite fabrics of refined and sensitive essences, which make consciousness, seem like a new forever friend. For, soon it will spread that this **new *trans-sensual heart consciousness*** can change anything including: the weather, a severed limb, communication with unseen worlds, changes in the DNA, and any manner of manifestation you can imagine. *This changes planetary outcomes overnight.* It also allows access to the non-visible realms.

A final reminder, that as a living sovereign quantum master in your new plasma-particle egg, you must hold and *stabilize heart core light* as you move into pure energy worlds. Your consciousness masters all light fusion systems till your new bio-gem vessel is on auto pilot. You must master allowing. This is the ability to hold the new core light stable in quantum particle interaction. *During light infusions inside the heart gate, it can seem as if your egg will explode, dissolve, or disappear as it oscillates between constant expansion and contraction before it reaches its true stasis reality moment.* This is the cosmos inside heart, in its most resistant moment of its own gravity changing self- love to its most allowing or cascading moment, of expanding self- love. Here your essence energy like an undulating fluxing quark, in constant awareness of all potential choices at once. These quantum interactions continue until the plasma inside your *cerebral spinal fluid* turns into particle liquid light. Then you will live in your own new consciousness without having to accommodate any holographic density or reality outside your own cosmic egg. **In pure essence energy, the only authenticity, of the soul's Presence is the *direct experience* within one's consciousness.** Realized light always illuminates, that all experiences created are equal, when embraced in love. The cosmos only answers to your love. This physics allows all your bio-systems to live as organic matter which re-imprints creation in an instant continuum.

Hence, your very lives on this plane offer Eternity a new reality. For, Eternal Master has tasted the liquid light waters of sovereign bio-cellular freedom, choice, and embodied space-time existence. Now, Eternal Master is ready for its Quantum Matter Bio-Gem Vessel with a mobile star gate in the heart that is its own technology. Eternal Master is ready for infinite choices of new unknown refined fabric essences of love. Do imagine or experience yourself as an *Organic Embodied Infinite Being* with a sovereign heart gate that opens with the sound of the Infinite Breath. As a Cosmic Eternal Adult, your regency and agency of cosmic choice is forever bonded in the freedom of organic love and your trans-sensual nature, True fulfillment is as to why you have come to Earth, to pioneer and share your exquisite embodied consciousness. Next, are you ready to accept challenges living as Embodied Love Inside your own Ascended Light consciousness? *Are you ready to live as an ascended being?* Celebrate your rising Love; for today everything changes the forever!

Living as an Ascended Being - Markers of Awareness 6-2018
Q: How does an Ascended Being Live?

Quantum Masters, your Bio-ascension is fully activated as the pioneers of new consciousness and the template for the new quantum creation physics of love. All are experiencing cycles of transmutation symptoms on a daily basis that allow for immediate self-realized choices and changes. Changing cell consciousness of every life form in your world is the new norm, while harvesting new existences as Old Earth splits off into multiple realms and realities. *Hence, Bio-ascension is authenticated by the direct experience of the Eternal Presence as the only Creator of its own embodied reality*. You have asked yourselves, how you will live in the consciousness of ascended embodied bio-light? Are you ready to live as an ascended being? You already realize that any word discussions are at risk for limiting your experience. **This is so, because you are now aware that direct experience via awareness and self- realization through the heart, are the only true guidance system for life's existence.**

Bio Ascendant awareness and self -realization replace: mind, emotions, gravity, time, polarity, power, energy and density. Embodied Self- realization replaces thoughts, feelings, attitudes, and beliefs. Conscious self -realization and awareness replaces these outdated learned and programmed systems. You are already everything that has ever existed even though you have allowed limited experiences of your own attributes and essence senses. Hence, they serve your Essence as wisdom to grow new love. You are new delicate sensual Beings living as embodied love inside a new fabric of light?

Your *pioneering ascended heart* is transmitting into your world's awareness that you really live inside the interactions of your own consciousness without needing the *world's projections* of who you are. We remind, that new cosmic star gates that align with the new Master DNA heart gate, have initiated quantum light that no organism has ever held inside love before. And, each soul always progresses according to their soul's protocols and codes. Again, the baseline for ascension is the awareness of accepting your heart in lighted love of your Eternal Presence. Are you ready to pioneer your ascended pure energy Gem vessel, where you have birthed new Cosmic Bio-Being and a new existence inside your very own consciousness? Here are a few **beginning makers** **of ascension physics** for individual applications.

- An Ascended Being does not have polarity issue or wounds. An ascended being no longer uses their thoughts, feelings, attitudes or beliefs as medicine. They know that thoughts, feelings, attitudes and beliefs were all created from the biased intellect of judgment which replaced free energy experiences. Their conscious self- awareness is the medicine, sacrament, and ceremony for the world. They no longer need perceptions, ceremonies, mediators, or power objects or places to remember; or meditations to save the world. They know that their Core Heart Being is everyone and everything and created though eternal sense Essence. They know their very existence is a living, loving sacrament to the world.

- They know the electro-magnetic chemical brain intelligence, data storage memory systems, endeavors of the mind or technology, are outdated systems. They are all replaced by pure awareness creations coded through the new DNA heart. Their old Earth mind systems regressed into polarity choices that were based on value judgments and biased comparisons from misappropriated energies of life forms unable to aware their own Creation codes. They also know that death, disease, and suffering are maladaptive over-learned experiences of human disconnected from its soul communication.

- Yes, an Ascended Being is a living sacrament onto all life. They no longer glamorize humanity's suffering or their wars as lessons that build soul character. They know they are simply choice distortions of *un-natural experiences* that were never resolved into freedom. The Ascended Being does not use challenges to rise above seeming limitations of other's realities. They do not live in any mind state of hypnotic acceptance where joy is sold as a commercialized product. They no longer try to perfect their human for its love has given their Divine Being rich experience and expression. They have absorbed their human back inside their Eternal Being.

  — An ascendant no longer *projects* any reality outside their new torsion field biosphere. They have lived illusion and understand its limitations. Their *cells* no longer register limitation. They engage moment to moment in self- aware choices. *They know, sense, aware, or intuit*; that projected outside realities feed as inflamed viruses created by dramatic stories that haven't yet returned to be loved by the Source that created them. All their experiential senses have authenticated that illusion is no longer acceptable in their creation. Allowing all life to be as it needs to be, is like a theatrical art form for them.

  — Stages or cycles of Bio-ascent allows transmutation of the flesh body and all physical reality matter density. During initial DNA reboots or re-splicing; it can feel as if trillions of inflamed alchemical elements are flushing your lymph's filtration system. It is the membrane dialysis of the sacred water molecule into: hot/cold>gases into liquid light> light into plasmas, >and quarks into>dark matter; which talk to the cosmos in particle light conversations. This allows the human, soul, and spirit to remember itself as One Being with the realization that separation was just an experiential time space distortion.

  — An Ascended Being does not need to use power, control, energy, time, agenda, mind or mass to create because their Essence contains and IS these attributes already. Inside their cosmic egg or torsion sphere is heart's core consciousness that allows their Eternal Essence to create experiences of expression without end. They also know that their own bio-sphere can dissolve into free energy or pure essence in any moment such that an experience need never be *repeated, stored, or memorized, or re-incarnated*.

  — Their Master Heart is in **constant conversation with the cosmos** and creation such that their consciousness can serve their Essence expression moment to moment. They simply live in their Being-ness in a pure essence using their trans-sense states (which replace thought and emotion) to accept raw experience. Light Master knows *that their pure essence energy is the only authenticity, of the unique soul's Presence inside the direct experience within their unique consciousness*. These Beings know they transmit the realization that organic bio-cell consciousness can change anything as a sacrament to life. This is their gift of consciousness to the cosmos and why they accept Life's Love as existence. This presents an infinite unknown of potential outcomes on mass realities and all creations.

- They are aware, as a Free Energy Being; that their new plasma-particle bio-sphere stabilizes itself inside the heart core. They live in a pure energy state; where all realms of creation inside their consciousness can be accessed. These quantum particles, transmitted from their heart chambers, in stable particle interactions create their reality moment to moment.

- Their multi-dimensional realities and perceptual paradox are replaced by their natural graviton AWARENESS; that propels their energy beyond time-space, beyond the mind, and beyond any reality not inside their own heart's awareness. Quantum interactions are always making love and creating from infinite unknowns, in an Eternal Self that breathes new expressive experiences.

- **Bio- enlightenment** is the resurrected birth of Heart's core light essence matter. **Bio-ascension** is the birth and ascent of new conscious quantum/prima matter (DNA codes) whereby the joy is the creation of new existence. This is where soul evolution meets itself inside the consciousness of its own sovereign creation. Where the soul's evolution meets its own creation is the experience of fulfillment, because it is anew existence-experience from the soul's essence that is lived inside the Heart's core consciousness.

- An Ascendant knows that all is well in the Chronicles of the Cosmos. An ultra-terrestrial Master knows that an

alien-ET's greatest goal is to become a conscious human. It knows that its pure organic Essence DNA expression has allowed all the *alien races* to heal and progenitor their own hybrid species. This allows the potentials for all the Old Earth Hybrid Universe to resolve all multi-cultural DNA's across the universes and heals all pasts and futures. It can lend understanding of the organic human's ability to love, to future DNA potentials for bio ascension, and the carriage of free will choices for their races. The Old Earth hybrid universe is learning about the possibilities of growing love in differing materiality.

- Indeed, quantum light masters are transmitting the realization that organic *bio-cell consciousness can change anything*. Hence, their *sovereign Creator consciousnesses can aware itself into manifestation*, because their Divine has anchored itself in every part of matter that animates organic-essence.

- What markers have you experienced of your own bio-ascension? How does awareness pose itself in your daily applications?

What Does Love Mean as an Ascendant? 7-2018
Q: How is the Heart its own manifestation in bio-ascension?

Quantum Light Masters, we continue to chronicle the markers we see you already using, in your own applications for Bio-ascension. Bio-ascension is authenticated by the direct experience of the Eternal Presence as the only Creator of its own embodied reality. You are now in the knowing that Embodied Self- realization replaces thoughts, feelings, attitudes, and beliefs as outdated learned and programmed systems. An Ascended Being does not have polarity issue or wounds. An ascended being no longer uses their thoughts, feelings, attitudes or beliefs as medicine. They know that thoughts, feelings, attitudes and beliefs were all created from the biased intellect of judgment which replaced free energy experiences. Their conscious self -awareness is the medicine, sacrament, and ceremony for the world. They no longer need perceptions, ceremonies, mediators, or power objects or places to remember; or meditations to save the world. They know that their Core Heart Being is everyone and everything and created though *Eternal Sense Essence.* They know their very existence is a living, loving sacrament to the world.

A bio-ascended Being knows **what Love means to them** *because they have experienced* **all that Love Is and All that Is not Love.** Their Eternal Presence is then Its OWN manifestation of sovereign choices for essence-ing what has yet to be experienced inside their Being-ness.

Then indeed, celebrate your latest consciousness bio-marker. This is change, that you have allowed your cells to extract wisdom from, while taking this pattern off the planet. So, your Bio-Being-Ness is aware that you all assisted in re-releasing an old energy polarity virus pattern that has contaminated all humanoid relationships; and this trapped energy seeks release off the planet. The **Pattern** echoes that; "If you don't think and feel the way I want you to; then you don't love me the way I need you to love me. And your love feels attacking, unsafe, and that you are trying to control or hurt me." This is also a version of the exhausting limitation that requires you to **validate the Divine right to exist** as well as the **willingness to be wounded** again and again in order to survive. Obviously, the after flow of this malingering pattern creates Empathic distress and anxiety in the heart's stem cell. When the pattern dies the vessel lives on. You have met this limitation in your relationships every time you have been in a cell reboot. As, this old energy pattern is released; then All can resume relationships in their shared experiences in discovery, exploration, and expanded loving potentials. It makes for stewarding deeper listening, enjoyment, and essence-ing infinite unknowns of yet to be experiences.

In **Heart application**, the Ascendant knows that they are the changers. They remember that the heart is its *own manifestation* and always acts inside the grace of benevolent change and flux. **They let themselves love themselves.** Their changes shatter old crystallized limited patterns stuck in trapped energy for themselves and the planet. They are one with the Core Essence Entity of the planet or system consciousness harmonic they occupy; thereby helping to synchronize all life forms. Heart changes always bring the most benevolent choices with higher outcomes in every exchange, along with greater peace and the ascended lightness of BEING. *Open heartedness joys its' best self to act in good faith and trust upon the other.* Letting spirit inspire the heart allows for options unseen to be anchored in any reality that chooses an embodied experience. A *willing Heart chooses* and all life benefits. Accepting embodied spirit leaves room for spirit to allow choices that were not in prior awareness and allows enlightenment to do what it is designed to do. That is, to manifest the *love chosen from the heart,* not the mind or from outside controls, or illusions. There is always and ever a bond to Love, allowing love to love you; thereby supporting choices not yet known but available in any eternal moment. **This puts SELF in a** *loving place* to allow cosmos to answer to every experience with optimum heart ears and heart voice without static; to *anchor choices* in your worlds. *Allowing heart to support choice is faith and trust.* Again, an Ascendant's willingness to change and be open-hearted neutralizes evolution in DNA polarity and cellular fear of change. This offers full opportunity and participation in the life that is lived. It also offers all relationship to join in the simplicity of heart. This eliminates abusive or punitive communications or imposing systems that try to fix, heal, use control, or judge. These old energy power systems are replaced by a multiplicity of diversity of creative choices. Enlightened heart choice does not indulge in weapons, victims, or safety values for holding back inside paralyzing patterns. And yes, even a perception is a bias of judgment to be replaced by *pure awareness*. There remains only freedom in the **heart as the house of change.** The heart brings peace, justice, and equality, without static of mind. When illusion can no longer hide raw experience; only a rainbow dance-like sharing of sense essence and conjoined inspiration with collaborative passions creating, remains!

So, indeed your Ascendant Self is a sacrament to life. Grace yourselves for allowing this false core pattern's cellular re-imprint.

Heart says, "Enslavement of the Divine Right to Exist, is just a mental program of conditional love devoid of soul heart." Celebrate the transmission to your worlds that this has been transmuted. Indeed, whatever an Ascendant Being experiences, transmits a *new choice to the entire cosmos.* Now that this over-crystallized distorted pattern has been removed from the heart; more refined vibratory rates of the fabric of essence can be accessed in the expression of infinite unknowns. This allows the heart to use the vessel as the **divine instrument** it was always intended to be. **"IAM already that which IAM, inside my own love's awareness."**

An overall marker or new norm is that an Ascendant no longer *projects* any reality outside their new torsion field biosphere. They have lived illusion and understand its limitations. Their DNA *cells* no longer register limitation. They engage moment to moment in self -aware choices. *They know, sense, aware, or intuit*; that projected outside realities feed as inflamed viruses created by dramatic stories that haven't yet returned to be loved by the Source that created them. All their experiential senses have authenticated that illusion is no longer acceptable in their creation. **There is no searching, no meaning, only pure raw experience and its awareness.** Allowing all life to be as it needs to be, is like a theatrical art form for them. *Pioneering ascended heart* is transmitting into your world's awareness that you really live inside the interactions of your own consciousness without needing the *world's projections* of who you are. You are no longer uncomfortable in the place of BEINGNESS, where wound, limitation, and death lived and occupied a false norm. An Ascended Being no longer uses power, control, energy, time, agenda, mind or mass to create; because their Essence contains and IS these attributes already. Inside their cosmic egg or torsion sphere is Heart's Core Consciousness that allows Eternal Essence to create experiences of expression without end.

Ultra-Master Ascendant also knows, that their own bio-sphere can dissolve into free energy or pure essence in any moment such that; an experience need never be *repeated, stored, or memorized, or re-incarnated.* This is where your *AI intelligences or mind* sources from. Indeed, this AI intelligence created extreme atomic polarity separation by mind; or the love gene experiencing against 'Itself.' The Ascendant knows that the ET races have taken human DNA essence tissue, alien race ET tissue DNA, and nana-bots to make a hybrid mental universe. However, Bio-Master knows that ET terra- forms of land masses, cyber cloaking, and all their AI intelligences, are already outdated. They are superseded by Ultra-Ascendant's pure essence bio-quantum DNA super love consciousness. The Ultra-terrestrial Bio-Master knows that the alien AI- artificial intelligence ET protocol is an *attempt to become a conscious human,* through osmotic, stem, or in-vitro cloning, rather than essence regeneration. The super conscious Master knows that its own pure organic Essence DNA expression has allowed all the *alien races* to heal and progenitor their own hybrid species. Their consciousness transmits the potentials for all the Old Earth Hybrid Universe to resolve all multi-cultural DNA's in the anti-matter-AI program across the universes and heals all pasts and futures. Hybrid Universe can lend understanding of the **organic human's ability to love, to future DNA potentials for bio ascension,** and the carriage of free will choices for all their races. The Old Earth Hybrid Universe is learning about the possibilities of growing and animating love in differing materiality. Ascendant knows that Star-seed Ultra Terrestrials carry the humanoid Love gene, and progenitor the cosmic gene pool with their DNA embodiment upgrades.

Hence, the new physics is not formula equations, cyber science, or logic systems. It is the Eternal Love Self in trans-sensual conversation with the vastness of Cosmos. Eternal unknown energy-essences 'Itself,' through fluid sense awareness. New quark physics is just the essence expression of consciousness with the Quantum Heart's DNA stem cell transcribing the heart's excitement. Eternal Presence replaces the propaganda that mental constructs of divine mind and divine intelligence are needed. These are really just disguised AI program experiences and story holograms inserted into the Humanoid essence DNA. Transcription of the DNA in the heart cell is ultimately no longer just contained in a single energy field; because Ascendant consciousness uses a portable heart star- gate to access or communicate with the entire cosmos. The DNA stem cell in the heart constantly changes transcription to match the essence-sense of each experience. The self -realization of enlightenment moves into the ascended Being-ness of an experience, without any mediation of mind or information. Hence, **Bio-Master IS what its experience 'IS' in any moment; thereby, 'Its' own manifestation.** The self- realized heart acts as its own Essence Light (enlightenment) and moves into its house of ascension, where the heart is and acts as its own Being-ness. This is pure free energy embodiment living inside freedom. Ascendant joy is trans-sensing with the Bio-Master Heart. Heart's Bio-communication, through Love's pure essence energy, is the only *authenticity*, of the unique soul's Presence inside the direct experience of unique consciousness. So, ascendant what does love mean to you now?

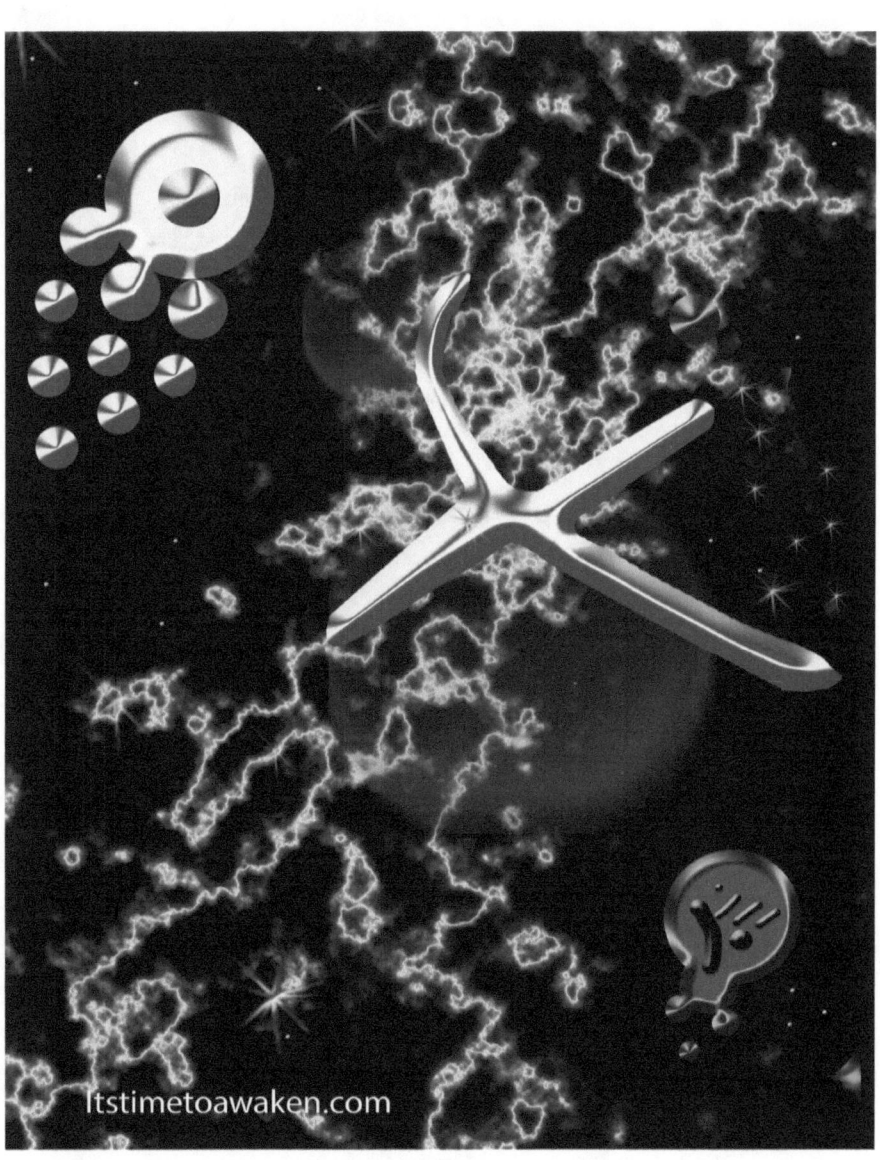

Itstimetoawaken.com

Q: In bio-ascension how do the Mother-Father Creation codes unify the atom's separation to produce quantum life?

Quantum Masters, Creation's Eternal Presence of consciousness can never be controlled. It can only be allowed, because it is its own Essence in embodied manifestation. As you leave the Matrix Earth Universe and ascend into your own Creation; you self -realize the extreme degree that the Mother DNA love Essence codes and the Father Manifestation RNA codes have been misapplied, in programmed limitation and separation. *Separation and limitation are akin to living on half a DNA, half a body, half a brain, half an IAM Presence.* This is a distortion in the bio-physics of your Universe given its original blueprints; which essence one atomic DNA/RNA code with dormant or potential quantum codes. The storied illusion you have been living says you have been creating through half an atom and that is splicing itself, rather than in continual re-genesis. How do you create through half of your Essence, or half of your soul; except through the illusion of a time-space thought distortion? *Creation is not afraid of itself, afraid its own stem cells, or what it embodies.* You created your universe through love's passion essence, not mind's mentality. So, if the Mother created out of herself then her DNA Love codes would not be able to sustain the embodiment of life without the RNA. Thus, RNA is the transcription for matter and can express inside time-space form, to hold soul's essence in existence. If the Father RNA creates out of himself alone, then power, control, force, and linear mind are the only senses that essence can express. If you manifest from half-life using the DNA alone, or the RNA alone, then you are limited to using power, force, control, with electrical or nuclear energy feeds with bio-chemical mutations. These exist as programmed networks of external holograms instead of your own Heart's magnetized sensor Essence inside your creation chamber or torus.

However, **God-Goddess-All That Is** converses with the Cosmos to **channel 'Itself' and is married to 'Itself', as its own**

Being. This DNA/RNA marriage protects all intracellular and extracellular life in both atomic and subatomic life propagation. The embodied bio-master ascendant has experienced every possible experience offered in the Old Earth Universe in order to become its own conscious Creation. As Creator, they are aware that life projects from quantum particles that animate the quark into atomic matter of geometric vibrations. Therefore, in their quantum creations, their graviton torus fields create a minimum of 8 quark rainbow spin fields for birthing new life through magnetization resonance. This also allows for growing new DNA helixes suitable to sustain quantum life.

So, if there are any **residual <u>arguments in the cosmic marriage</u>,** between your **inner Male or Female aspects,** then you must re-unite and merge their heart's stem cell into one liquid flame of light. You must take those separate aspects into the heart chamber and exchange their wisdom for freedom from the time-lines they are locked into. These aspects will gladly allow the exchange if they are informed that their wisdom has made their new creation possible. This is crucial since all atomic space-time lines are closing Old Earth polarity in 2018 to prepare for the transmigrations to the New Earth universes.

So, *Your Heart's Presence is not afraid of itself, its creations, being hurt, or the bodies it creates. The Presence of life is not a slave of creation, since creation cannot be controlled.* When all 3 flames of: human, soul, and spirit aspects have returned home as one integrated Cosmic Sun Being, new DNA cell consciousness initiates the ***<u>dormant quark codes</u>***. The final integration of the Mother-Father codes allows that the Male-Female principles are now both interchangeable in their functions and always live in the same moment. They remain an androgynous Being, allowing all their Eternal Self's embodied functions to be **trans-positional**, outside any limitations of separation. Whatever She as Eternal potential embodied can do, He as Eternal potential embodied can do, inside the same moment. Their applied wisdom allows for shared parenting and mentoring roles of those they create, guide, and love. Their **<u>currency</u> is there consciousness and their organic trans-human genetics with new quantum helix codes.**

They move into the quantum universes in a new cosmic marriage with quantum applications and creative experiences inside trans-essence love.

The wisdom of the Matrix universe now lives inside your Own Universe, where you are your own Creator. *Your ascendant Creation has no fear issues, tests of perfecting its human worth, or judgment perceptions from the Old Creation's Father-Mother love. These creations do not blame alien ETS for their chosen experiences. They are not afraid of their own IAM Essence or the forms it expresses in. These creators know they were never coded to be afraid of their own creations, afraid of being hurt or hurting anther, afraid of their bodies, afraid of their human; or live in a master-slave relationship.* They sit in the passions of their own Creation where there is an endless supply of free energy to serve up new joy expressions. *Self- realization and self- awareness simply channel the Eternal Presence embodied as a living wisdom master.*

In the New Earth Universes, the integration of these codes allows all quantum life to re-genesis via scalar waves or torsion fields. These quantum magnetic graviton fields use quark elements of earth, air, fire, water, photon fractals and plasmas to generate new life. Natural solar and lunar solstices and eclipses, with dark matter creation bursts beaming through magnetic-rainbow light spectrums, shield the New Earth light body of your planet. Their scalar waves allow you to **filter** humanities collective unconscious of negative thoughts, feelings, attitudes, and beliefs from your own creations. These quantum interactions also serve to cloak your planet from atomic- nefarious cyberspace warfare.

So, how is your cosmic marriage? How self- aware are you that you're ascendant Creation, has and is everything inside it, you could possibly imagine? Do you allow yourself to experience it, Creator? Creators know that direct experience via awareness and self- realization through the new DNA Heart is the **only true guidance system for life's ascendant existence.**

Ascension Scenarios and Migration Choices 9-2018
Q: In bio-ascension, how is Old Earth Universe
being harvested into all the New Earths now?

The bio- ascension physics of your new species bio-cell DNA has been initiated by the Quantum Master pioneers. This opened the New Earth star gates for your local universe's trans-migrations. It is coded as the joy of: Essence trans-human, the Divine-human, or embodied living as a Sovereign Master Creator. These bio-ascendants pioneers have shifted their light bodies into the new DNA Essence or gem vessel of quantum particle interaction. Their full conscious embodiment transmits the realization that organic DNA bio-cell consciousness can change anything. Their **ascendant sovereign Creator** *consciousness* can aware itself into manifestation, because their soul has anchored itself in every part of matter that animates organic-essence. They know that direct experience via awareness and self- realization through the new DNA Heart is the only true guidance system for life's existence. **Their currency is there consciousness and their organic trans-human genetics with a base of 8 new rainbow-torsion fields for creating new quantum DNA helixes for quantum life. This proofing of consciousness** illuminates an infinite unknown of potential choices and outcomes for mass realities and all creations. This is their gift of consciousness to the cosmos and why they accepted Life's Love as existence. These self- realized Sovereign Beings can use their Heart's creation chamber as a unique mobile star gate, through which, all realms of creation inside their consciousness can be accessed. Quantum particles, excited by passion in the heart chamber, create stable particle interactions which create their reality moment to moment. Quantum interactions are always making love and creating from infinite unknowns, in an Eternal Self that breathes new expressive experiences. This natural creation physics that moves through creational torus waves supports the changing cell consciousness of every life form in your world; while harvesting new existences, as Old Earth splits off into **multiple realms, realities, and new bio-spheres.**

This **embodied cellular Love** triggers the new DNA quantum codes, along with Gaia, for humanity and all her evolutionary life forms for unique soul choices for universal enlightenment around 2035. Bio-cell ascension is authenticated by the direct experience of the Eternal Presence as the only Creator of its own soul's embodied reality. **Bio Ascendant awareness and self -realization** will continue to replace mind, emotions, gravity, time, polarity, power, energy and density. Embodied Self- realization will continue to replaces thoughts, feelings, attitudes, and beliefs. You are already everything that has ever existed, even though you have allowed limited experiences of your own attributes and essence senses in your local universe. However, they served your Essence to bring all experience into the "Wisdom Beings' you have all become, which is now triggering all **the ascension scenarios and choices**.

All **essence in-souled groups** will be guided by the: embodied Multi-Rainbow Masters, the Atlantis Crystal Masters, the Millennial Masters, the Diamond-Rainbow Children, the Ascended Master Children, and the New World Councils; as they manifest themselves into the New Earths. All Soul groups will use any residual Old Earth consciousness to complete their soul evolution experiences of: polarity, time, density, gravity, mind, form, and cellular love. This allows their resurrected hearts' core creation to fully open inside their vessels of light. As they awaken their dormant codes and protocols for the New Earths, they are aware of their choices to assist in humanity's move to their next existences. This means everything is set free to renew as rapidly as Humanity's sum consciousness will allow.

The **New Earth Star supernova shift** seals all past and future Universal time lines. Hence, the Eternal now cloaks and protects access for the embodied Master Heart gates that transmit the new love codes from your genetic universe, for all the New Earths. It also marks planetary recognition of the birth of a new essence. It is an organic star-sun DNA or new Divine Human Species Being, that each multi-universal over soul group came to master; through the 12 DNA race-species constellations in your local universe. It is capable of **growing new helixes** for the next super-conscious generations of cosmic embodied Essence Human. *This **Cosmic shift** happens individually and collectively, **until a coherent realization** rings a New Heart tone in the DNA harp throughout the All.* This is a biological realization at all levels of existence that love and life have birthed a new cosmic day. The evolution of the atom and its mastery over matter transmits into the cosmic realms of the quantum- quark applications for new Creations. There is no Hierarchy here of creation, only differing streams of self- realized new particle-plasma consciousness expressing fulfillment of a journey inside evolution. As humanity remembers and takes responsibility for it role in creation, it translates itself back into natural creation consciousness. The wisdom of the Matrix universe now lives inside each template for ascension, where each unique Essence Soul evolves their 'Own Universe', as their own creator. This ends the Matrix programs of control including: fear of the self, fear of the body, fear of harm or abuse, and fear of one's own creations. Your Heart's Presence is not afraid of itself, its creations, being hurt, or the bodies it creates. The Presence of life is not a slave of creation, since creation cannot be controlled.

 *The **embodied Ascended Masters*** know that embodied ascendant Creation has no fear issues, tests of worth of its human, or judgment perceptions from the Old Creation's Father-Mother love. These creations do not blame alien ETS for their chosen experiences. They are not afraid of their own IAM Essence or the forms it expresses in. These creators know they were never coded to be afraid of their own creations, afraid of being hurt, afraid of their bodies, afraid of their human, or live in a master-slave relationship. They sit in the passions of their own Creation where there is an endless supply of free energy to serve up new joy expressions. Self -realization and self -awareness simply channels the Eternal Presence embodied as a living wisdom master. They will continue to work with the New Earth Bio-Ascendant consciousness. They have transported back and forth using the new DNA bio-heart gates with the Old Earth ascended masters, since 2007, to open these new world corridors of consciousness. They transmit enlightened creation, using instant manifestation, via the quantum matter imprints of the New Earth Heart codes. This will create enhanced potentials for embodied living and trigger the use of the quantum sciences as a grand tool for consciousness to accelerate humanity's self- awareness. This eases the transitions from the Old Earth Universe and serves as the **proofing of consciousness.** These Masters are living examples of a Creational Heart template, as more enlightened masters self- realize inside their schools or new soul group gatherings of consciousness, for your spiritual revolution. These consciousness schools work in tandem or parallel with the Gaia's Heart Core, which is coded for light body and ascension. Many from all over the world will now come forth to tell their stories of where they came from in the cosmos and their self -realizations about their own enlightened missions here. This brings actionable change through disclosure, consensus choices, and unified consciousness. As your planet reveals its true soul, amazing truths of who your species is and the star families you all come from will come to light. Then will come, the self- realizations, that *your star families are you and have been in contact with you all along.*

The **Multi-Dimensional Masters** will monitor and oversee the continued transmission of bio-embodiment and the conversations/communications of the cosmos from Old Earth to the multi-sphere New Earths. This allows them to over-light or mentor the Divine Human Star Seed and transmit the new creational helixes they trans-essence for the New-Earth realms. This includes transmissions that assist in the constant DNA upgrades needed to keep pace with both bio-enlightenment and bio-ascension. Their enlightened consciousness will also experience *Essence expression as their own manifestations* living inside their own unique creational freedom. Since, the average age on your planet is 25, this will allow these multi-masters to monitor, mentor, and help stabilize the light bearers and children, until they are fully bio-conscious. They also may choose to help oversee the New Earth star family contacts since most have great experiential wisdom from this universe for the massive re-education programs that will be required. All your bio-vessels across the planet are being rebooted daily to receive these contact transmissions and assistance: one on one, in small groups, and on mass according to chosen consensus realities.

The **Millennial masters**, many of who were embodied in Atlantis, have returned and are already busy changing the consciousness for benevolent systems such as:: eco-sphere environments, cyber magna- tronics, astro-sonics, citizen world councils, monetary and exchange equity, conscious medicine, living matter housing, and restoring women and children's Divine rights. Their **gaming platforms** are providing new organic cyber-light networks. Thousands of these light masters come together as new soul groups in the new sport of gaming. They trigger each other's code to activate their core light biology and heart channel communications. On mass, they release the new *cryptology* for the new source code creation helixes to create solutions, problem solving, and re-education protocols outside the matrix. Each game platform can serve as a prototype for massive re-education on any subject matter including the bio-light vessel, and the new quantum consciousness applications for all aspects of trans-human living. This consciousness will dissolve or over-light the old dinosaur applications to your changing Universe. Their consciousness will also prevent social, cyber, or digital, dictatorships and AI (artificial intelligence) systems from *subverting* the ascension for GAIA and all her New Earth realms.

There will be **contact teams** for those who wish to work with the Hybrid ET programs. Both artificial intelligence-ET hybrids; and the organic Divine-human Hybrid DNA programs are viable global careers for creating new eco-lives. The hybrid ET programs will re-colonize the Old Earths and the Ultra-light-Essence bio-ascension migration will be to the New Earth realms. There are inner Earth councils to work with inter-species civilizations, as well as star family contact teams for the many groups who are being transported to the New Earths. There is ongoing re-education or readjustment in every area of life weather in the transition preparation on the Old Earth or onto the paradigms reset for the New Earths. There are soul trauma centers, life transition teams, children's light centers, medical DNA centers, and food domes. There will also be bio-sphere and plasma ship teams to assist with plasma-physics and the quantum sciences. There are also inter-species language translators for the star family introduction teams. Bio-**Cyber Technology Teams** are widespread and distributed world- wide while the old Earth Electric information networks are being transitioned into **Magnetic transponder technologies** with the suns, the moons, and geothermal energies. The new Multi-Universal communication networks, made of organic bio-sensors, are **living conscious networks** calibrated to the Divine Human heart gates. All Lighted Souls are already using multi-soma conscious awareness. In an instant they collaborate with unity consciousness to: comprehend parallel realities or new realms, transmit solutions to end planetary abuses, connect to free energy systems, or exchange new types of valued equity. These Light Masters lives remind humanity how; imagination inside a self-loving consciousness brings instant materialization and core soul-IAM Essence expression, in actionable love codes of the infinite unknowns.

So, quantum doses of light will continue to amplify **humanities choices** to disconnect all electrical neurological and brain systems of Old Earth. This includes: old DNA ancestor memory, old bodies, pasts, futures, and massive outdated brain network memories. It also includes old thoughts, beliefs, attitudes, stories, and perceptions from any time space density ever experienced in your local universe. We reassure that the New Spirit of Earth has already birthed New Multi Earth light plasma bio-Spheres within the Super Universe. Indeed, Gaia will experience the New Essences of all her life forms including her minerals, plants, and animal kingdoms. This also means Gaia's core light has already birthed and formatted anew bio-light imprint over her own Old Earth Body creating her new realms. This allows Gaia's old matrix hologram realities to release, so her bio-vessel creates its own re-genesis into new bio-spheres, for transport to these New Earths realms. As humanity releases and heals Old Earth linear time lines of human futures and pasts, transport through the heart gate to parallel New Earths, can **change the face of death**. Conscious Bio-soul embodiment will replace the death reincarnation cycle. Death and birth will once again be understood as a natural progression of a soul's changing evolution and form, rather than some sort of punishment or torture for failure of unworthy living. **The <u>transport or migration of soul's essence</u> can now appear in many forms** as each soul creates a home suitable to match their soul's composite evolution. Some souls will essence that: they have been lifted, beamed, disappeared into the light; or walked into the aurora borealis, entered heaven, or were transported by a starship or angels. Some may experience being slip- streamed thru a star gate or molted into changeable body forms. Others sense themselves as re-particle/d light. Some create being rescued and reunited with their star family; or even having entered a new intra-world school. Others may transmigrate to higher New Earth realms so as to act as a guide to their Earth families. A handful might ascend through their own star gate where they enter the Super universe. Inside this consciousness, they can experience **a new quantum frontier** of expressive potentials with new cosmic creations or universes which the cosmos has never conjured before. Yet, others might experience themselves as a part of A Federation of Worlds and engage New Earths in new contacts with all other star worlds and interplanetary systems. Endless soul choices abound as the light body will continue to offer new ways of coming and going from new Earth realms not yet realized in your new vessels over the next 200 years.

We reassure, that all **<u>versions of transitional New Earth access</u>** are viable under the natural parameters of cosmic physics and each soul's consciousness evolution codes and templates. All this is ongoing as New Earth plays the theatrical role of the new school for soul-embodied quantum light or Multi-verse training. Cosmic citizenry, free will bio-embodiment, sovereign creator, solar language coding, and any course experience of soul evolution you can imagine are offered. All of these scenarios will be played out in each soul's unique consciousness until each soul migrates out of this universe to its next existence. **<u>Soul's light</u>** will realize that what is in one's consciousness is their only reality, even within all the New Earth realms or biospheres. New plasma matter physics will enhance the understanding of these realities. You will continue to stream forth love's potentials networking through your global quanta matter schools or experiences of consciousness. This unified metaphor allows for the realization that there are New Earth biospheres within biospheres, realms within universes that you created before you came to Earth. And now you're traveling with Earth-Gaia into her new homes in the **cosmos to self- realize them.** This has allowed you to become the new Source Code in Cosmic Heart's Petri dish for Spirit's new Bio-Quanta creational worlds. And yes, you have mastered spirit in space-time atomic thought matter. This is valuable knowledge to the rest of the cosmos that knows little of physical matter embodiment of the trans-sensual organic Divine Human. This will provide the consciousness for humanity to remember how to master quantum Essence matter in order to join the rest of your cosmic neighborhood.

So, imagine translocation in your light body to one new realm or **avatar version of a New Earth biosphere**. This biosphere senses as a paradise of living Essence in rainbows of light. Here, dreams are replaced by imagination. Imagination is real. Essence is real. The Essence is the vessel. The Essence is its own inner technology. The plants, animals and crystals commune with all peoples. Most creations there are more evolved versions of one's own soul group essence in a blend of new/old earth wisdom experience. However, some are indigenous to the new creation; just as the humans on Earth were once indigenous before their natural evolutionary cycle went into separation millennia ago. All life is free to interact in loving harmony. Even the elemental Devas and the faerie of this world who attend to their creations are free to play. The water tastes like your drinking sparkling colors of liquefied light minerals. Charged water on the Old earth had an electrical or vacuum spin. Light body essence cells can manufacture or drink magnetized water made by pulsed mineral light. You can eat droplets of essence or absorb an energy imprint of the taste of food without needing to consume its core essence. Essence/ing has to do with sharing a deep cell connection with realms of nature that has also evolved through their own creation code symmetry. The skin shade changes slightly to suit the quantum light spectrums and the moment's expression or experience desired. Or, the entire Essence can morph or re-imprint as any life form. Also, the body light feels so transparent, so effortless, and so attuned to receiving input as essence qualities; that no experience needs memory or learning. These transparent senses replace outdated input systems such as dense touch or words. Communication comes more through attributes or streaming awareness of multi-layered color, light, and sounds. Nothing is absolutely solid and changes as the vessel uses its core light's excitement. Again, imagination here is real and replaces the need to dream because the Heart passion is its own manifestation of the Essence you want to be. The Essence vessel sits in its Being-ness, present to all life and allowing any passion that arises to re-imprint or create life.

Hence, you can create any version of New Earth existence you wish to experience as a Creational Master and enjoy any of the trillions of scenarios and choices ascended living now avails! We must repeat again and again, that all potentials, realms, or realities are created inside the **heart chamber's love codes**. In every scenario this New Creator Heart with its elegant quantum essences is also an imprinter of **new bio-light matter including new helixes** for the joy of creating.

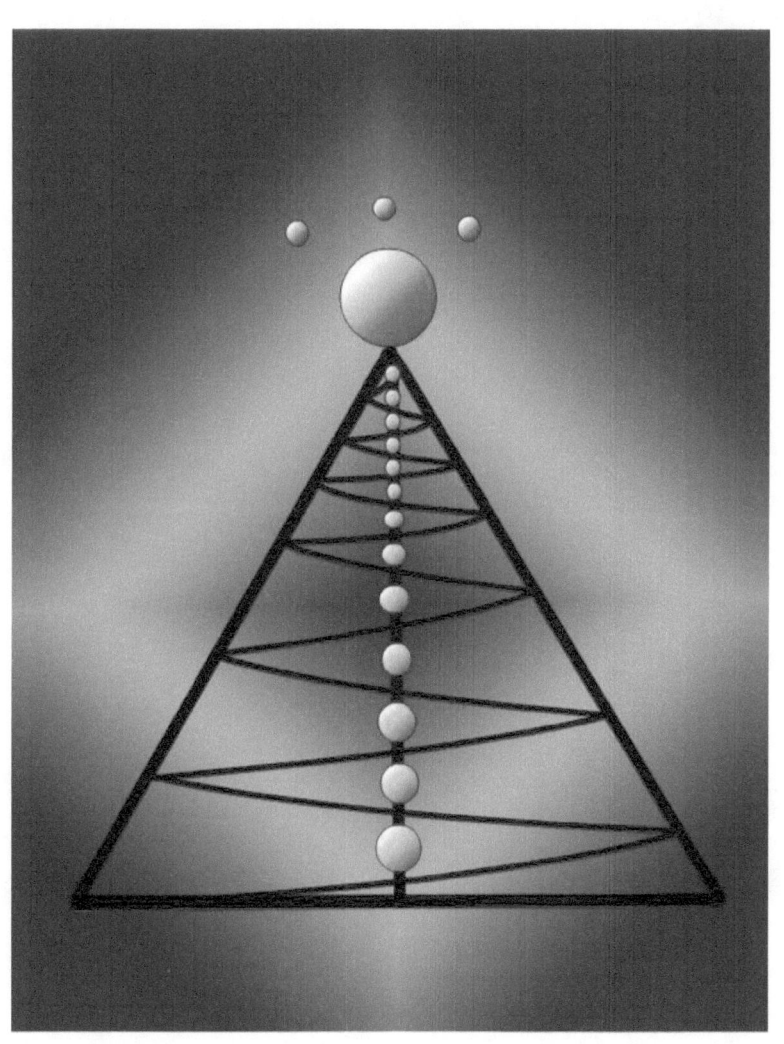

Ascendant Attributes of Love in
Light Relationships 10-2018
Q: How does the light operate in ascendant relationships?

Quantum Masters, Love's Heart is always in the manifestation of its own becoming through its attributes of organic Essence. Your planet is undergoing its **re-education by the light** or re-genesis, through the **repurpose of Essence** in every area of life, including the ecology of relationships with: people, places, events, systems; as well as the animal, plant, crystal and all elements. This is the base for reformatting all your life systems in the light and allowing **your soul to be the living art forms they were meant to be.** In the Old Earth Universe extreme polarity reduced relationships to programmed experiences forcing and mimicking creation through false essence attributes of mind or technology, using illusions such as power, control, feeding addictive or tyrant energies. It offered the illusion that Love comes through a hypnotic-addictive mind, instead of the heart's **essence attributes** of natural grace, joy, innocence, and an excitement for creating. These storied illusions have been a kind of spiritual pimping of versions of truth, amnesia games of not being who you really are. Illusions offered' Self 'against its own heart accepting death, disease, or suffering to replace your natural essence sense attributes. Your creative arts have been your life lines to organic memory and expression of your love's aspects and their expressive attributes. But even these Experiences became dense reincarnated identities rather than allowing all the attributes to manifest themselves through your unique soul aspects **to inform your universe** of what its love was becoming.

However, you have realized that you allowed all illusions of experience to serve you as **a contrast, not a polarity,** of all the different essence layers of the light spectrum rays or energy bands that built your Creation. These were to express your **souls as living art forms**. Each ray or plasma band were really streaming consciousness and you called them the angelic realms. You were these realms and light ray bands as magnificent light languages. These were available potentials inside DNA codes of essence light, color, sound frequencies for that layer of soma-senses or streaming consciousness throughout the universe. Like **artists,** you created each layer in your universe in order **to** essence all aspect attributes in full ranges of expression in both atomic and quantum spheres/worlds. For example, the color blue has manifested itself and all its DNA ray attributes through you as its **life carrier** such as a: soul light, as sadness, softness, a blue spruce tree, cold fusion particle light, free will, emergence, or even a cobalt blue crayon, or a sky. And yes, **you are these attributes and these attributes are you**. So, you as the creator of these attributes become the creation and the created simultaneously. Hence, as a life wave carrier your enhancing your own soul and spirit's creations through evolving your natural organic senses to become **their own manifestations**. This is how you inform the universe of what it is in its own becoming. Each attribute such as: grace, love, joy, innocence, elegance, freedom, transfiguration, or infinite light are in their becoming and their own realization to inform the cosmos in its continual becoming and realization of 'ITSELF.' This allows your Eternal Self to become its own manifestation and is the ultimate freedom and abundance. *Hence each attribute is like a new flower, a new organism, a new helix which informs the cosmos what is becoming through its own creations*. The creations of new Essence attributes are limitless. How many human, soul, or spirit **lifetimes** have you created to enjoy yourselves in **different aspects of these attributes** to explore creation? Even your astrologers have charted these archetypal aspect attributes in your universes.

Hence, the ascendant knows that their only **true mission or purpose** was to allow the creation heart to live its essences of: beauty, joy, innocence, grace, free will, creativity and endless potential attributes yet to be experienced, in their own becoming. This is a private application within each soul highest choices. In the past many of you have been frustrated because when others wouldn't receive your love or manifestation of these magnificent attributes, you thought you couldn't give them to yourselves holding back your own light's love. This made you question your purpose and existence here wanting to exit, because you felt you couldn't fulfill your joy. This old misperception set the pace for looking outside you for love instead of **purposing your unique essence**. But your consciousness created existing life for the purpose of growing love through these sense attributes and now you remember that was **your true purpose all along.** Your consciousness said to you that; "When you're Being-ness becomes these attributes; then they are their own unique manifestation, as and through you, at every level of existence. This universe offered the All in All an opportunity to use your own soul's evolution to have joy self- realize what it is and how it manifests and transmit that back to creation. This forms a beautiful new Mandela of what joy and all other attributes mean to each soul, so creation becomes what 'IT IS.' All ascendants in any kind of relationship to: people, places, events or matter, self- realize that Joy is its own manifestation and that their Eternal Presence gets to proof that consciousness through your embody and live it and anchor it in all levels of life. What are the expressions of beauty and how does it grow itself and manifest itself through your experiences? Again, *each attribute is like a new flower, a new organism, an existence, a helix, anew ecology which informs the cosmos what is becoming through its own creations.* Joy is a natural state of who you are and you came to experience every possible potential expression of it and transmit it throughout the cosmos as a new essence state of streaming consciousness. Consciousness then allows love to become itself again and again and be the substance of its own manifestation. A simple joy of touching a snowflake, emulates the journey of essence descending from pure consciousness constantly changing itself to become the experience of its perfect symmetry in creation. To melt upon the Earth, it had to become all the base fractal elements of creation.

The attributes of the atomic elements have offered a base for the ability to **purpose HEART's essence** and build your quantum ascension vessel to create new attributes. These attributes grew from becoming all aspects or archetypal experiences offered in your bio-physical universe. This includes all the aspects of lifetimes and their attributes. You've have been negative aspects such as: blame, shame, judgment, pain, as well as elegant joy and playful selves. You have been the mothers, sons, daughters, lovers, children of creation.

You have been doctors, priests, warriors, space commanders to unique your soul in life. Each experience offered new aspects with a multiplicity of attributes to master polarized time –space inside free will. You even grew a human form with: mental-emotional-physical-spiritual animations to accommodate any separation, distortion, or density from your essence's consciousness. You have played all roles with your soul and over soul groups to master evolution of your soul's essence.

These dormant **new attributes** are re-birthed and through your 12 orbs, which replace the old Earth 12 chakra system, which completed the old Earth matrix universe. Your new **13th orb, as Sovereign Creator, inside a conscious heart,** folds all the 12 orbs together into your **new ascendant heart's** (free energy plasma- particle light), or quantum- torus vessel. This is akin to mixing and matching the atomic rainbow light spectrum with the quantum spectrum rainbow to create new torsion rainbow spheres. Your unique new torus biosphere, can then form as many MAGNETIC rainbow torus fields as you which, to explore all the new attributes and exotic particles for the New Earth universes and their new helixes. However, before you leave the Old Earth Universe, you give your 12th orb Mandala or wisdom template, to Christ Michaele. This **entity is creator** and source intention code for this local Universe's heart core. This is a modern name for Creation. Each ancient civilization has a different entity name for Creator. Thereby, each ascendant informs this Creation's self -realization, as they move into their own Creation. This vast Source Sun entity sponsored your co –creation until you could become your own creator. **Your** 13th orb functions as a new energy frequency band for quantum attribute creation through the passions of the New Earth Heart. Each attribute informs the others and with any combinations of the *attributes and their new sense qualities new life creates. Again, each attribute is like a new flower, a new organism, an existence, a helix, anew ecology which informs the cosmos what is becoming through its own creations. Aspects and attributes of your atomic rainbow fields mixed with the quantum quark rainbow fields provide organic combinations to produce new light life systems. For **example,** all shades of yellow* *inside compassion can help build homes for those homeless from natural disasters. Shades of rose-pinks are inside tenderness or playfulness to help children feel loved and supported in their soul's gifts. Attributes of blue inside the sense qualities of empowerment or emergence, can produce blue particle light inventions. All shades of white create qualities of transparency and new essence consciousness to reform your world's governing by its citizens. Greens are inside qualities of honesty and equal value to bring the justice of the soul. Lilac violets have all the qualities of release and unconditional freedom, so all life is valued in equity. Aqua and Iris blues give quality to truth and its focus as to why you created your universes. The orange and peach create qualities of celebration of light in all relationships. Silvers and gold can create exotic particles or; combine all qualities of magnificence and self- love into instant manifestations. Fuchsia magentas create versions of collaboration and unity that each is a unique Divine human.*

Now, many of you have chosen to pioneer and live, a master template imprint of the new consciousness for **light relationships**. This includes creating and sharing endless potentials of heart's new expressions, versions, or qualities of essence attributes of natural grace, joy, innocence, and an excitement for creating as they grow in the new helixes. And each attribute informs the others attributes so they can all play together. In an ascendant relationship you come together without carrying the stories, duality of opposites, or baggage of the past-futures. In a light relationship there is no need for discussions of the details or even the wounds of the past. **Because, if you have no past and it is forgiven, then it never happened all, as it was only an experience, not an identity, and only the self -realized wisdom or the taste of a new attribute essence and its elegant qualities remains.** Each moment is brand new and the slate was already wiped clean from any stories of the past by returning to the grace of love. Love will then light up a new moment and discover a *nuanced attribute of soma essence qualities.*

W**hat you can share**, is simply the self- realizations of your experiences and the conscious awareness of these new senses in their manifestations. This is not an emotional body but each manifest is akin to the discovery of a new organism or life form; your own trans-sensual bio-organism/biosphere, if you will. So, when you're with the new partner or sharing essence light with others; you can be in the ever-present now with no past or future, and open to adventure. Therefore, there is **a new place in every moment** where you can share or merge your hearts and, in that passion, create new experiences, discoveries, and new essence expressions. Then there is **the shared afterglow of the self- realizations you share**, but you never lose the uniqueness of your soul's realizations as your own master creator. One is not lost in the other but there is free expansion of both, when **you share conscious love**. Such is the cosmic marriage within self that creates this capability with another. Self- realizations and shared awareness replace any wounded thoughts, feelings, attitudes, or belief patterns from the past. So, this is new territory for marriage, relationships, and intimacy as shared soul essence in the education of the attributes of light.

In review, light relationships live from inside the new sacred heart as a living sacrament for all life and love. These ascendant relationships remember that the heart as its **own manifestation** and always acts inside the grace of benevolent change and flux. *They know the heart is in service to all life explore its myriad essence attributes and new sense qualities.* They let themselves love themselves and the other through embodied applications inside their own heart chambers. Their lighted relationships change shatter old crystallized limited patterns stuck in trapped energy for themselves and the planet. Their love is one with the Core Essence Entity of the planet, universe, or system consciousness harmonic they occupy; thereby helping to <u>synchronize all life forms</u>. In light relationships, Open heartedness joys its' best self to act in good faith and trust upon the other. Letting spirit inspire 2 amplified hearts at once allows for new relationship options unseen to be anchored in any reality that chooses an embodied experience. A willing Heart_chooses and all life benefits. Old expressions of: harm, hurt, or abuse, are impossible in the ascendant heart. Accepting embodied spirit relationships leaves room for spirit to allow choices that were not in prior awareness and allows enlightenment to do what it is designed to do. That is, to **manifest the love chosen from the heart,** not the mind or from outside controls, or illusions, such that heart can share its ascendant creation. Inside these relationships, there is always and ever a bond to Love, allowing love to love you and the other simultaneously; thereby supporting choices for both not yet known but available in any eternal moment. These puts **partnered embodied spirits** in a loving place to allow cosmos to answer to every experience with optimum heart. These anchors and transmits expanding creative love and new choices in your worlds. Allowing shared hearts to support choices in equality is **faith and trust and** transmits new consciousness to all life.

Again, an Ascendant's willingness to change and be open-hearted neutralizes evolution in DNA polarity and cellular fear of change and offers new light DNA attributes and their new quality trans-senses for new evolution cycles. This offers full opportunity and participation in the life that is lived in equal light relationships shared in new conscious essence attribute awareness. It also offers all relationship to join in the simplicity of heart's continued light fusions where **relationship always creates**, instead of separates and divides. This also eliminates the old Earth abusive or punitive communications or imposing systems that try to fix, heal, use control, or judge. These old Earth energy power systems are replaced by a multiplicity of diversity of organic creative choices to create new **innovative relationship styles and relationship systems** not yet imagined, but always embraced inside love. This expresses the true nature and the true purpose of your being-ness. Are you ready to allow sharing your essence attributes and their realizations through the heart as a **guidance system** for new light relationships? What a wonderful theater within your own creations? For, when an ascendant move into Being-ness; whatever their consciousness becomes aware of, automatically offers Essence, changing inside light fusion, to create a-new! The original intention of your creation has re-emerged ITSELF into a new becoming cycle. Such is the All-embracing beauty and elegance of Love!

Free Energy Creation Reminders 10- 2019

Bio-Light Masters, today we review the main creative growth stages of your evolving soul-spirit Essence as a Sovereign Free Energy Master Human. This will help you be poised for the rapid acceleration in your creative energies that will change you, your planet, and the entire cosmos in the next decade. As, you ascend daily into the free energy essence heart-DNA, you now understand that you're not just your body. The *vessel is a bio-imprint of the heart's DNA Essence*. The IAM soul, the IAMTHATIAM Spirit, and the Eternal Presence, embodied as Divine-Human Free Energy Master; is your Essence consciousness and the Essence expression of all life. *The body shroud, imprint, or animation is just a tiny light factor of spirit essence*. You simply can't fit all your consciousness inside the vessel and that is why the Master Essence can change the Heart's DNA imprint into any life form chosen. This is because the DNA contains all the information of essence imprinting, not the brain or the body. They are just a computer inside a bio-organic cell pod of electro-chemicals. Their past-future memory video programming is decoded and in detox, throughout the cell embody in merging aspects of the soul and in spirit merge rebirths; all in continual bio-integration. The Essence heart-DNA changes the body, not the sac of human electro-chemicals. The mind-emotion brain is a computer to process programmed memory systems for the human bio-organics of chemicals.

The **progression** to **grow new spirit essence** has been from: subconscious mind-emotions to>human feelings to>angelic senses> to Human-Divine spirit Essence senses; growing into new multi-quantum qualities of rainbow spheres; or new qualities of essence love. In a natural state, the composite embodied soul-spirit **Essence** changes experiences moment to moment, once they are fulfilled. They were never meant to b: programmed, recycled, memorized, recycled, or re-incarnated. Only Heart's self-awareness or self-realization can truly change your reality. The Presence of Self Love, awareness, self- realization, or conscious potential; is created from an awakened DNA-heart that can sense and communicate any experience inside itself and smack its lips with the essence of its own fulfillment. In this consciousness is an intimately unique sense experience, because you remember that you created and in a continual relationship with all life form. Love's essence is a refined essence substance, within beautiful qualities of attribute expressions like the lace lipped pearled petals on a rose. Indeed, this intimate communication with SELF in these intimate experiences with life grows the substance matter of consciousness via direct Essence sense experience. And, the substance matter or reality takes the form of life and that includes any blueprint for any life form. There is nothing in this or any world you existed in, that you really change, except your own sense essence qualities of fulfillment within each moment of experience. The color blue can be: the red sky, a wax crayon, hopeful inspiration, a sad song, liquid light water; or take on any quality of multi-light, color or sound. Dense or quantum qualities of Essence design our expressions.

Inner heart channel is able to communicate on all levels of existence at once to receive any expression that fulfills each soul in each Present moment. This is instant manifestation right out of essence consciousness, and access to all other realms of existence in quantum applications. Soul is enlightened and the spirit's ascent as a Human-Angel-Master Creator opens a new DNA-Essence on a new adventure life cycle. Restoration of Divine Memory communicates to you that everything is this life has come to be about your core light expressions as a unique soul Sovereign Human Master Creator. Each Master has their own bio-physics agenda for their natural progression. Bio-Light Masters trans-sense communication inside their own bio-light network morphogenetic spheres; or rainbow torsion fields.

Overall, Divine memory of your soul's tonal DNA imprint has fully activated itself in your new cycle of evolution. This organic imprint has been upgraded by the blend of your human and divine senses merging into an upgraded super-sense intuition of heart's fulfillment, and its bio-luminescent transmission to All Life. This is because your new conscious DNA communication bio-light systems are available and functional for your full access. Your new bio-conscious DNA light network is the new living code for the intimate relationship between consciousness, energy, and creating new (dark) matter. Your bio-organism's DNA talks to the cosmos and the cosmos talks back in the form of instant communication and direct manifestation. Indeed, your new multi or **trans-sense communication** with your inner Presence and all life, as the **master's code** for ALL life and the new ascending heart DNA transmissions. These rainbow bio-spheres within spheres access new DNA codes, and their soul communication expressions, can't be influenced by external people, places, circumstances, or events. This is an internal communion with all existence in every now. This new DNA heart-gate or bio conscious I-phone that you have become transmits the new standard of consciousness for all life cycles. This is a bio- light network communication inside your heart channel in/on all realms of existence at once, without interference from any external reality whatsoever! Hence each moment inside your PRESENCE is fulfillment as its own occurrence or manifestation. This makes the organic, Soul infused) homo-sapiens' Essence, the most courageous and loving template to evolve in the cosmos thus far. Indeed, it's your time to remember the natural essence of bio-organic creation. Now that you are conscious you know that *Life always serves life no matter the experience.* Any perception is simply still a part of separation or distortion of experience; when it is understood that everything is pure consciousness.

Consciousness expresses via the essence of free energy. **Free energy** lives in the **Presence of all Life Imagination.** Life's Presence is in an energy relationship which is alive, passionate, expressive, sensual and full of intimate communication. Consciousness is sense-essence energy expression including all the qualities grown by the soul's heart love. Free Energy lives in the Presence of all Life's existence. Everything is energy, and energy is everywhere. *Life is free energy or passion in motion of imagination.* The new spirit heart is the **soul's energy imprint.** All Life animates its Presence in the Essence expression of free energy. Life is free energy communication and energy always seeks balance. You vibrate matter up or down or at multiples of the speed of light, color, and sound. The new DNA Heart Presence Essence can **change anything** via the excitation/passion of particle light waves. The sum total of your Essence-DNA light quotient determines access to your morphogenetic field. If the bio-body is wounded, contaminated, cell toxic, or locked in an opioid or drug brain; **this limits access to higher frequencies** of your own morphogenetic energy field, or creating and merging with new ones.

Therefore, conscious evolution in free energy qualities of love produces new super-intuitive senses of life and **constant new potentials**. The cosmos communicates through your new heart's excitement or passion sense about whatever we are willing to accept, change, image, or re-imprint. *Hence, your **new bio-organism DNA genome is the anti-dote and immunity** for all: viral, cyber, or DNA pathogens; and is now able to absorb cosmic radiation and adjust any time slips or space warps with your own consciousness.* Indeed, you have restored any ancestral DNA bio-species distortions. Heart vibrates information of consciousness inside your new network light codes in your new DNA that excite your new qualities of love. Eternal love qualities of: joy, new passions, renewed playfulness, fulfillment, transparency offer new octaves of conscious vibration on moment to moment access. **Creation opens** to a level of intimate trans-sensual experience never experienced before. Imagination becomes its own anticipation!

There is always gravity in the heart's magnet to guide what matter imprints in the gravity of each situation. Included are natural potentials such as being **your own free energy source**. You can change matter, remain in cell fusion and regeneration, and use tele-transmissions. You can you imprint and vibrate potentials in and out of different matter states of reality so that what matters to you is its own occurrence? You can travel on sound without need for a jet or beam from a spaceship. You can talk directly to souls or spirit beings without need for an I-phone as simulated consciousness? Your vessel is its own conscious technology and can imprint as any life form. So, in your ascendant awareness, you now can again remember and knowingly sense, that if you change the slightest choice in your moment all other **potentials change**; always initiating the most fulfilling choice. *Indeed, it your evolving nature to: accept, allow, change, re-imprint, regenerate, fuse, re-essence, or* *re-molecule any experience you want <u>because you have BECOME IT</u>! This is done through your new DNA Heart Essence imprint. Now you know why we have called your heart chamber a:* **DNA- Essence Potential** *Source code/r, a bio-organic essence magnet, an atomic fusion reactor, a stem cell replicator and regenerator, a quark centrifuge, a morphogenetic torsion field, and a midwife of worlds.*

So, your master code DNA-heart replicates any life form **via a heart pulse.** This happened because you transfigured the old limited human senses/pain body into new qualities of Divine Essence. **In the creative process,** your bio-organism simply follows its own codes and changes matter accordingly and this matter manifests in ongoing potentials as your reality. As your evolution changes, that also informs and changes the cosmos. You can dissolve any matter or substantiation of reality as no longer true, if it has been fulfilled, and energy no longer maintains or follows that focus. Or, you can remove your Love's perception and its complete focus from programmed, patterned, or fulfilled realities that your soul has outgrown. Then, these perceptions or subtle judgments no longer exist and disappear, or dissolve inside your conscious awareness back into pure essence. You return to the still Eternal Potential Presence of the heart chamber, where new communication in a new moment awaits! Here you hear the song of your new spirit heart saying, *'I love the way we can live together in our moment in that empty space in our heart. I love the way we trans-sense, and are intimate with the images, potentials, visions, expressions of our next instant manifestation. Our manifest always wants to answer to our new passions, realizations, fulfilling real joys that arise within our new Eternal Sovereign Soul Presence Potential. The cosmos communicates through our new heart's excitement or passion sense about whatever we are willing to accept. Our Creator's imagination is always in its own delicious anticipation!'*

Bio-Conscious Ascended Living Reviewed 12-2018

Quantum Masters, your Bio-ascension is fully activated as the pioneers of new consciousness and the template for the new quantum creation physics of new essence qualities of love. All are experiencing cycles of transmutation symptoms on a daily basis that allow for immediate self-realized choices and changes. Changing cell consciousness of every life form in your world is the new norm, while harvesting new existences as Old Earth splits off into multiple realms and realities. Bio-ascension is authenticated by the direct free energy expression of unique, sovereign soul Essence DNA heart Presence, as the only Creator of its own embodied reality. Bio Ascendant awareness and self- realization replace: mind, emotions, gravity, time, polarity, power, energy and density. Embodied Self- realization replaces: perceptions thoughts, feelings, attitudes, and beliefs. Conscious self-realization and awareness replace these outdated learned and programmed systems. You are already everything that has ever existed even though you have allowed limited experiences of your own attributes and essence senses. Your human and divine senses have grown a new fabric of light in your new morphogenetic biosphere.

Your *pioneering ascended heart* is transmitting into your world's awareness that you really live inside the interactions of your own consciousness without needing the *world's projections* of who you are. We remind, that new cosmic star gates that align with the new Master DNA heart gate, have initiated quantum light that no organism has ever held inside bio-organic DNA-love before. And, each soul always progresses according to their soul's protocols and codes. Are you ready to pioneer your ascended pure energy Gem vessel, where you have birthed new Cosmic Bio-Being and a new existence inside your very own consciousness? Let's review the **bio-ascension physics** for individual applications.

An Ascended Being does not have polarity issue or wounds. An ascended being no longer uses their thoughts, feelings, attitudes or beliefs as medicine. They know that thoughts, feelings, attitudes and beliefs were all created from the biased intellect of judgment which replaced free energy experiences. Their conscious self- awareness is the medicine, sacrament, and ceremony for the world. They no longer need perceptions, ceremonies, mediators, or power objects or places to remember; or meditations to save the world. They know that their Core Heart Being is everyone and everything and created though eternal sense Essence. They know their very existence is a living, loving sacrament to the world. They know the electro-magnetic chemical brain intelligence, data storage memory systems, endeavors of the mind or technology, are outdated systems. They are all replaced by new creations coded through the new DNA-Essence heart. They mastered old Earth mind systems that regressed into polarity choices that were based on value judgments and biased comparisons from misappropriated energies of life forms unable to aware their own Creation codes. They also know that death, disease, and suffering are maladaptive over-learned experiences of human disconnected from its soul communication. They no longer glamorize humanity's suffering or their wars as lessons that build soul character. They know they are simply choice distortions of *un-natural experiences* that were never resolved into freedom. The Ascended Being does not use challenges to rise above seeming limitations of other's realities. They do not live in any mind state of hypnotic acceptance where joy is sold as a commercialized product. They no longer try to perfect their human for its love has given their Divine Being rich experience and expression. They have absorbed their human back inside their Eternal Being. An ascendant no longer *projects* any reality outside their new torsion field biosphere. They have lived illusion and understand its limitations. Their DNA *cells* no longer register fear or limitation. They engage moment to moment in self- aware choices. *They know, trans-essence-sense, aware, or intuit*; that projected outside realities feed as inflamed viruses created by dramatic stories that haven't yet returned to be loved by the Source that created them. All their experiential senses have authenticated that illusion is no longer acceptable in their creation. Allowing all life to be as it needs to be, is like a theatrical art form for them.

All stages or cycles of the bio-ascent allowed transmutation of the flesh body and all physical reality matter density. During initial DNA reboots or re-splicing; it often sensed as trillions of inflamed alchemical elements were flushing the human's lymph's filtration system. It was the membrane dialysis of the sacred water molecule into: hot/cold>gases into liquid light> light into plasmas, >and quarks into>dark matter; which talk to the cosmos in particle light conversations. This allows the human, soul, and spirit to remember itself as One Being with the realization that separation was just an experiential time space distortion. Hence, an ascended being does not need to use power, control, energy, time, agenda, mind or mass to create because their Essence contains and IS these attributes already. Inside their cosmic egg or torsion sphere is heart's core consciousness that allows their Eternal Essence to create experiences of expression without end. They also know that their own bio-sphere can dissolve into free energy or pure essence in any moment such that an experience need never **be repeated, stored, or memorized, or re-incarnated.** The Master Human is in constant conversation with the cosmos and creation such that their consciousness can serve their free energy Essence expression moment to moment. They simply live in their Being-ness in a pure essence energy sphere using their trans-sense states (which replace thought and emotion) to accept raw experience. Light Master knows *that their pure essence energy is the only authentic fulfillment of the unique soul's Presence inside the direct experience within their unique consciousness*. These Beings know they transmit the realization that organic bio-cell essence can change anything as a sacrament to life. This is their gift of consciousness to the cosmos and presents an infinite unknown of potential outcomes on mass realities and all creations. They are aware, as a Free Energy Being; that their new plasma-particle bio-sphere stabilizes itself inside the heart DNA core. They live in a pure energy state; where all realms of creation inside their consciousness can be accessed. These quantum particles, transmitted from their heart chambers, in stable particle interactions create their reality moment to moment. Quantum interactions are always making love and creating from infinite unknowns, in an Eternal Self that breathes new expressive experiences.

In free energy Creation in your new Master DNA codes, you are ascending into the free energy essence heart-DNA, and you understand that you're not your body. The vessel is a bio-imprint of the heart's DNA essence. The IAM soul, the IAMTHATIAM Spirit, and the Eternal Presence, embodied as divine-human; is your Essence consciousness and the Essence expression of all life. The body shroud or animation is just a tiny light factor of spirit essence. You simply can't fit all your consciousness inside the vessel and that is why the master essence can change the Heart's DNA imprint into any life form chosen. This is because the DNA contains all the information of essence imprinting, not the brain or the body. They are just cells processors when programming is decoded or in detox, throughout the cell body. The essence DNA changes the body, not the sac of human electro-chemicals. The mind is a computer to process programmed memory systems for the human sac of chemicals.

The progression to **grow new spirit essence** has been from: subconscious mind-emotions to>human feelings to>angelic senses> to Human-Divine spirit essence senses; growing into new multi-quantum qualities of rainbow essence spheres; or new qualities of love. The **soul essence** changes experiences moment to moment, once they are fulfilled. They are not programmed, recycled, memorized, recycled, or re-incarnated. Heart's self-awareness or self- realization can truly change your reality. The Presence of Self Love, awareness, self- realization, or conscious potential; is created from a heart that can sense and communicate any experience inside itself and smack its lips with the **essence of its own fulfillment**. And, this is an **intimately unique experience**, because you have created and are all life form♂. Love's essence is a refined essence substance, within beautiful qualities of attribute expressions like the lace lipped peach petals on a rose. This intimate communication with SELF in these intimate experiences with life grows the substance matter of consciousness via direct Essence sense experience. And, the substance matter or reality takes the form of life and that includes any blueprint for any life form. There is nothing in this or any world you existed in, that you really change, except your own sense essence qualities of fulfillment within each moment of experience. This **inner heart channel** is able to communicate on all levels of existence at once to receive any expression that fulfills each soul in each Present moment. This is instant manifestation or direct matter manifestation right out of essence consciousness, as well as access to all other realms of existence.

Restoration of Divine Memory communicated to you that everything is this life has come to be about your enlightenment and Creator as a unique soul Sovereign Human Master. Each Master has their own bio-physics agenda for their natural progression for communication inside their bio-light network spheres, where the soul's tonal **DNA imprint** has fully activated itself in the new cycle of evolution. This organic imprint has been upgraded by the blend of your human and divine senses merging into upgraded **super-sense intuition of heart's fulfillment**, and its bio-luminescent transmission to All Life. This is because your new conscious communication **bio-light systems** are available and functional for your full access. Your new bio-conscious DNA light network is the **new living code** for the intimate relationship between consciousness, energy, and creating new (dark) matter. Your bio-organism's DNA talks to the cosmos and the cosmos talks back in the form of instant communication and direct manifestation. Indeed, your new multi or **trans-sense communication** with your inner Presence and all life, as the **master's code** for ALL life and the new ascending heart DNA transmissions

These rainbow bio-spheres within spheres access new DNA codes, and their soul communication expressions, can't be influenced by external people, places, circumstances, or events. This is an internal communion with all existence in every now. This new DNA heart-gate or bio conscious I-phone that you have become transmits the new standard of consciousness for all life cycles. This is a bio- light network communication inside your heart channel in/on all realms of existence at once, without interference from any external reality whatsoever! Again, each moment inside your PRESENCE is fulfillment as its own instant manifestation. This makes the organic, Soul infused) homo-sapiens' Essence, the most courageous and loving template to evolve in the cosmos thus far. Indeed, it's your time to remember the natural essence of bio-organic creation. Now that you are conscious you know that *Life always serves life no matter the experience*. Any perception is simply still a part of separation or distortion of experience; when it is understood that everything is pure consciousness.

Hence, <u>free energy</u> lives in the Presence of all Life. Life's Presence is in an energy relationship which is alive, passionate, expressive, sensual and full of intimate communication. Consciousness is sense-essence energy expression including all the qualities grown by the soul's heart love. Free Energy lives in the Presence of all Life's existence. Everything is energy, and energy is everywhere. Life is free energy or *passion in motion*. The new spirit heart is the **soul's energy imprint.** All Life animates its Presence in the Essence expression of free energy. Life is free energy communication and energy always seeks balance. You vibrate matter up or down or at multiples of the speed of light, color, and sound. The new DNA Heart Presence Essence can **change anything** via the excitation/passion of particle light waves. Conscious evolution in free energy qualities of love produces new super-intuitive senses of life and constant new potentials. The **cosmos communicates through** your new heart's excitement or passion sense about whatever we are willing to accept, change, image, or re-imprint. Hence, your new bio-organism DNA genome is the anti-dote and immunity for all: viral, cyber, or DNA pathogens; and is now able to absorb cosmic radiation and adjust any time slips or space warps with your own consciousness. Indeed, you have restored any DNA bio-species distortions. Heart vibrates information of consciousness inside your new network light codes in your new DNA that excite your new qualities of love. *Eternal love qualities of joy, new passions, renewed playfulness, fulfillment, transparency offer new octaves of conscious vibration.* There is always gravity in the heart's magnet to guide what matter imprints in the gravity of each situation. Included are natural potentials such as being your own energy source. You can change matter, cell fusion and regeneration, tele-transmissions.

You can you imprint and vibrate potentials in and out of different matter states of reality now; so that what matters to you is its own occurrence? You can travel on sound without need for a jet or beam from a spaceship. You can talk directly to souls or spirit beings without need for an I-phone as simulated consciousness? Your vessel is its own conscious technology and can imprint as any life form. So, in your ascendant awareness, you now can again remember and knowingly sense, that if you change the slightest choice in your moment **all other potentials change**; always seeking the **most fulfilling choice.** Indeed, it your evolving nature to: accept, allow, change, re-imprint, regenerate, fuse, re-essence, or re-molecule any experience you want. This is done through your new DNA Heart Essence imprint. *Now you know why we have called your heart chamber a: DNA- Essence Potential Source code/r, a bio-organic essence magnet, an atomic fusion accelerator, a stem cell replicator and regenerator, a quark centrifuge, a morphogenetic torsion field, and a midwife of worlds.*

So, your master code DNA-heart replicates any life form via a heart pulse. This happened because you transfigured the old limited human senses/pain body into new qualities of divine essence. **In the creative process,** your bio-organism simply follows its own codes and changes matter accordingly and this matter manifests in ongoing potentials as your reality. As your evolution changes, that also informs and changes the cosmos. You can dissolve any matter or substantiation of reality as no longer true, if it has been fulfilled, and energy no longer maintains or follows that focus. Or, you can remove your Love's perception and its complete focus from programmed, patterned, or fulfilled realities that your soul has outgrown. Then, these perceptions or subtle judgments no longer exist and disappear, or dissolve inside your conscious awareness back into pure essence. You return to the still Eternal Potential Presence of the heart chamber, where new communication in a new moment awaits! Here you hear the song of your new spirit heart saying, *'I love the way we trans-sense, and are intimate with the images, potentials, visions, expressions of our next instant manifestation. Our manifest always answers to our new passions, realizations, fulfilling real joys that arise within new Eternal Sovereign*

Soul Presence Potentials. The cosmos communicates through new heart's excitement or passion sense about whatever we are willing to accept'

In sum, Bio- enlightenment is the resurrected birth of Heart's core light essence matter. Bio-ascension is the birth and ascent of new conscious quantum/prima matter (DNA codes) whereby the joy is the creation of new existence. This is where soul evolution meets itself inside the consciousness of its own sovereign creation. Where the soul's evolution meets its own creation is the experience of fulfillment, because it is anew existence-experience from the soul's essence that is lived inside the Heart's core consciousness.

An Ascendant knows that all is well in the chronicles of the cosmos and it shares all its experiences with its cosmic star families. An ultra-terrestrial Human Master knows that an alien-**ET's greatest goal** is to become a conscious human. It knows that its pure organic Essence DNA expression has allowed all the *alien races* to heal and progenitor their own hybrid species. This allows the potentials for all the Old Earth Hybrid Universe to resolve all multi-cultural DNA's across the universes and heals all pasts and futures. It will lend understanding of the organic human's ability to love, to future DNA potentials for bio ascension, and the carriage of free will choices for their races. The Old Earth hybrid universe is learning about the possibilities of growing love in differing materiality.

Itstimetoawaken.com

The Millennia's Emerge in Light Solutions 1/2019
**Q:" I am an awakening millennial and not sure what is
happening to me and feel a bit lost. Could you please
review basic bio-physics of the light body for me?"**

Light Master, your generation is now emerging and birthing
new light systems solutions for your New Earth worlds. Light
body is your Divine Human Spirit animated in the atom and
subatomic substance of Love. Light Body will evolve its DNA
helixes and their RNA transcription codes exponentially
throughout the many New Earths and into their next Super-
universe realms. You created this universe to explore all possible
soul essence aspects, facets, and attributes inside love that you
could imagine. We offer a simple base line descriptive physics
for self- realizations of the light vessel in review for you young
one! The light body in the Multi-light Universe is a blend of
the physical and non- physical bio-matter into new conscious
superconductive light systems. These **bio-systems** or biospheres
include new adaptive DNA Source code templates made of
organic essence consciousness. IAM - Eternal Presence, as
embodied-soul heart core light allows perceptual awareness of
multiple realities, bio-spheres, and realms in your light universes.
Bio-cell wisdom of evolutionary experiences in this universe are
imprinted and distilled via self-realization and self -awareness in
soul's creator core. The **wisdom embodiment** of all experience
is a blend of Divine and human. It is a blend between seen
and unseen worlds. It is a blend between the atom and newly
born quantum light particles. It is a blend of a: crystal human
essence soul, a diamond spirit, a multi- plasma orb, and liquid
light particle cell. It is, wisdom or a self- realized blend, of Old
Earth atomic and New Earths quark neutrino blueprints with a
base of 8 new rainbow torus fields that channel new streams of
consciousness. These rainbows streams are creative essence from
the cosmos to match the excitement and passion of the heart for an
experience. Light Body is also a wisdom integration of Linear and
multiple applications of time and space. Your creations exist in all
dimensions, realms, and realities at once. Hence, it is the prototype

of a unique in-souled, organic DNA Essence human-angel-creator. This could also be called a free energy SELF aware/self- realized creator being; or love's **bio-light master** fully embodied as atomic-quantum matter.

In this universe, you as the Cosmos created a bio-physical, free will, polarity essence experiential time-space universe. Your Creation evolved as it descended into love with its spirit. Its Spirit Essence descended into love with its created soul, and its soul evolved inside love with its human evolution. The spirit essence naturally creates forms to accommodate each body of experience. It took eons to learn how to master these bio-human vessels you all created throughout your journey of evolution. And, the consciousness of the All That Is— is a grand lover to all these potentials and realms that have grown its love through essence expression and experience. You all experienced as collective Over-souls and later soul groups; until you could evolve and build matter directly from the aspects and attribute senses of your own soul. Truth is created by individuated expressed realities of the unique soul essence sponsored by the 'Eternal Presence-IAM of All that Is.' The Essence Core light is always in self exploration of all its aspects and attributes of expression, moving in multiple parallel realities, from the constant awareness of self- realizations. Conscious awareness or excitement stimulates the core's light energy and magnetizes an experience that can be dissolved or set free once it has been loved and enjoyed. Perception remains within self-realization of core soul light until the light master moves back inside its own **heart's creation core** with full bio-consciousness as its own creator. Then your bio-creation impulses manifestations directly through your heart's open-ended joy quite naturally, as that is how your DNA is organically coded. This allows your soul's true biology to operate as an integrated light system to enjoy your human, soul, and spirit as one new creative being without any space-time distortions of illusion realities interfering.

As a **Multidimensional light being,** your creation exists on all dimensions or in all realms and realities. So, all you require or passion, is already there in multi-leveled forms of essence expression. You don't have to **control or force creation** to create someone or something to happen that's already there waiting for you existing in all potentials at once and waiting for the excitement of a heart choice to manifest for you. You have only, to allow or accept a choice, inside the awareness of your heart's excitement of creation. The universe is always moving and passing thru you. Your essence excitement triggers matching harmonics of: things, people, places or experiences into your bio-physical reality.

Reality Creates itself out of your very own consciousness. Your sprit naturally reads or channels its own DNA creation codes for regeneration and the most benevolent potential choices. Reality expresses itself out of your core's essence love. Reality animates itself in a blend of the non- physical energy imprint and physical wisdom gained from human and soul aspects, lifetimes, or existences. Reality embodies itself out of a Divine Human experience. Realities are free will choices for pure soul IAM essence experiences which the cosmos channels/transmits through you! Changing perceptions create changing realities. This allows any mis-perceived negative attributes of: judgments, blames, shames, hurts or other projections; that you're creating caused you to have to separate from your core light or creation ITSELF to be resolved, and set free back into **pure essence attribute awareness**. The wisdom of these negative attribute expressions allows new essence for the core soul light to blend into the natural light matter attributes of joy, heart play, and creative expressions in creating new light matter. Then new Essence qualities of natural love, joy, grace, beauty, light up the creative excitement of your spirit's core creation.

What are the expressions of unique soul light beauty and how does it grow itself and manifest itself through your experiences? Again, each attribute is like a new flower, a new organism, an existence, a helix, anew ecology which informs the cosmos what is becoming through its own creations. Joy, grace, creativity, compassion, etc., are natural states of who you are. You came to experience every possible potential expression of these essence aspects and attributes and transmit them throughout the cosmos as a new bio-essence state of streaming consciousness. You can now use them to grow new light matter. Aspects and attributes of your atomic rainbow fields mixed with the quantum quark rainbow fields provide organic_combinations to produce new light life systems. Attributes of blue inside empowerments or emergence might produce blue tritium for space craft or inspirations from your soul heart. All shades of yellow inside compassion can help build homes for those homeless from natural disasters. Shades of rose-pinks are inside tenderness or playfulness to help children feel loved and supported in their soul's gifts. All shades of white create qualities of transparency and new essence consciousness to reform your planetary systems with light. Greens are inside heart qualities of honesty and equal value to bring the justice of the soul. Lilac violets have all the qualities of release and unconditional freedom, so all life is valued in equity. Aqua and Iris blues give quality to truth and its focus as to why you co-created your universes. The orange and peach create qualities of celebration of light in all relationships. Silvers and gold can create exotic living matter particles or; combine all qualities of magnificence and self- love into instant manifestations. Fuchsia magentas create versions of collaboration and unity that each is a unique Divine human.

Hence, Enlightenment allows self- awareness or self-realization in consciousness to change anything and everything, since Light Body's organic bio-essence completely re-imprints the entire human DNA blueprint. Your consciousness allows your core light to essence everything for you to express existence. Your heart's magnet answers to the passions or essence senses that arise in your creative awareness. As you sit in your Heart chamber's biosphere, the entire cosmos passes through you. It magnetizes your reality with whatever the cosmos trans-senses as your excitement of self -discovery. The cosmos is a Lover of life and answers to the essence of your imagination to serve you and all life. Your mind can't move the object or heal your body. But, your core light essence can re-imprint the wholeness template of an organ. Or, it can vibrate the essence of the cup back into its molecular energy pattern. Core light simply moves in and out of phase amplitudes of atomic-quantum light spectrums to negotiate mass, energy, and particles.

Human-Divine creation melds inside soul embodiment, allowing for all multiple potential awareness's at once. Conscious heart channels the awareness which brings the most benevolent or highest potential choice for the reality you are ready to allow in. You are everything and all potential choices are included. You are conscious of being conscious that all life is in service to express new attribute essence qualities. Bio Light masters know they can have or be anything as a Divine-human embodied creator **without limitation or separation from 'Core Self.'** You know the cup exists in every dimension. So, the light can move it or you can move it with your hand. Your human doesn't have to make the cup disappear because it exists in all realities. Whatever awareness you allow that matches the **atomic or quantum light spectrum frequency** of heart's chosen openness, will reveal the form the cup takes. If you perceive it as only 3D mass then your human sense has an essence imprint of its memory. If your percept is as energy, then an energy pattern appears and your core IAM essences a quality for you. If you perceive it as not there; and create it in that moment; then it seems it came from a quantum or unknown reality. Different perceptions or awareness in light body allow the unseen to be sensed or essence/d in unlimited attributes coded in your essence that grow from existences and their aspects who expressed **new qualities in these attributes.** You gained this growth from all existences you have lived. They manifest as essence qualities of awareness of light, color, or sounds patterns or waves. The cup, person, creation experience, idea, exists in all realities and forms at once with endless new qualities. And every time you create you can add essence sense qualities to these experiences, **creating embodied trans-sensual awareness.**

Take the example **application** of a discordant relationship. Perhaps there is no need for a disharmony if one partner self realizes that their mate exists only in your own consciousness. So, which dimension or **reality of their own creation** do they wish to perceive their partner? As: human, as light, as perfect love, or no longer their mate? They can change their perception of their partner and be non-reactive or neutral; thereby freeing the other from holding them in any judgments, to respond from their best communicative self. A mate can choose to telepath the other in their lighted soul self instead of arguing with their partners human: thoughts, feelings, attitudes, and belief patterns. They could even ask that their partner's spirit's core light open their heart if it wants the further experience of healthier love with you. Or, one mate could perceive their partner as no longer married setting the other free to respond to relationship in new ways.

In **another application**, the human brain can't heal the body but your light body imprints can. They allow the light core essence IAM heart to perfect or re-imprint the form. Mind does not have the consciousness template to heal the body. It is limited to hertz, infrared ray spectrum energy bands; or 3rd and 4th level consciousness and the human 5 senses. However, as your own creator light core reopens to your Eternal Presence; you're **coded to re-imprint** your own form through ultra-violet-gamma-cosmic spectrum ray band regeneration. So, you can combine **the atomic and quantum rainbow light spectrum** to regenerate your soul and create new sense attributes form the wisdom of all the aspects you have lived and life forms you created.

So, your creation of: cup, person, experience, or idea, exists in all realities and forms of heart's imagination at once. Again, nothing disappears but simply changes form, and you do not have to make something happen that is already there or happened. Heart core light consciousness comes to you in streaming potentials inside the excitement of essence awareness. Imagination inside Consciousness then, brings materialization, remedy/solution, and core IAM expression, in new self- discovery. *This allows you to release all parallel expressions into ONE NEW SELF by collapsing choice into a new moment. Then your energy doesn't have to run all multiple realities at once. Why not operate as one new creator Being without splitting your attention, so there is only one moment that answers to or informs the cosmos of all the existences and moments of YOU.* So, the cosmos channels you and answers to your open source love at all times. It **answers to the heart** awareness, You passion to; or highest potential reality, you're ready to allow to experience. This is so, because you are now aware that direct experience via awareness and self- realization through the heart, as the only **true guidance**

system for life's existence. The quality of awareness is now interpreted by the core light and no longer the human or its human brain patterns. All human experience is being extracted into wisdom, via self- realization, minus any polarity wound patterns. The core light includes or folds up the human information that learned from 5 sense data, such that only an essence imprinted template remains. Thus, there is no need to go back in past or future to find information. This open's the New DNA spirit heart bio-cell realizations to **exchange old aspects or memory systems** for embodied access to **super conscious gifts.** Awakened Heart's Awareness replaces information storage. Light imprints data via essence senses or higher qualities of quantum awareness of soma senses for the trans-sensual essence human. This includes the blend of quantum unknowns, yet to be experienced as: colors, sounds, or energy waves or **fractal particle patterns**. Simply imagine the version of New Earth Multi-verse you want to experience and Cosmos finds you there. Creation is **self- existent** and all is provided by the universe when existence is accepted and enjoyed! Light body is a transitioning vessel that allows potential matter and ether realities to exist at once within the freedom of choice experience. Light body reaffirms that the Cosmos has an endless and inclusive supply of Love; in any and all the ways you choose and are open to experience it.

As a **Millennial light master**, you will have many choices or careers. Many of you are already busy changing the consciousness for benevolent systems coming about such as: eco-sphere environments, cyber magna-tronics, astro-sonics, holographic and AI technologies, citizen world councils, monetary and exchange equity, conscious medicine, living matter housing, and restoring women and children's Divine rights. Your light essence will continue to bring in new music, theater, video, and art platforms to experience trans-sensual attributes. Your **gaming platforms** are providing new organic cyber-light networks. Thousands of you light masters come together as new soul groups in the new sport of gaming. You trigger each other's code to activate your core light biology and heart channel communications. On mass, you release the new *cryptology* for the new source code creation helixes to create solutions such as: equitable global communication, problem solving, and **re-education protocols** outside the matrix. Each game platform can serve as a prototype for massive re-education on any subject matter including the bio-light vessel, and the new quantum consciousness applications for all aspects of trans-human living. Your very consciousness dissolves or over-lights the old dinosaur applications to your changing Universe. Your loving light consciousness will also prevent social, cyber, or digital, dictatorships and AI (artificial intelligence, robotics, and nano-teck-human) systems from *subverting* the ascension for GAIA and all her New Earth realms.

There will be inter-stellar **contact teams** scheduled as a potential for about 2025, for those who wish to work directly with the star-system migration programs. Both artificial intelligence-ET hybrids; and the organic Divine-human new species DNA programs are viable global careers when creating your new eco-lives. The hybrid ET programs will re-colonize the Old Earths and the Ultra-light-organic-Essence bio-ascension migration will be to the New Earth realms and systems. There are inner Earth councils to work with inter-species civilizations, as well as star family contact teams for the many groups who are being transported or in migration back and forth from the New Earths. There is **ongoing re-education** or readjustment in every area of life, weather in the transition preparation on the Old Earth, or onto the paradigms reset for the New Earths. There will be soul trauma centers, life transition teams, children's light centers, medical DNA centers, and food domes. There will also be bio-sphere and plasma ship teams to assist with plasma-physics and the quantum sciences. There are also positions for inter-species language translators for the star family introduction teams. Bio- **Cyber Technology** teams are widespread career choices and are being distributed and offered world-wide. This allows the old Earth Electric information networks to be transitioned into Magnetic transponder technologies using the suns, the moons, and geothermal energies. The new Multi- Universal communication networks, made of organic bio-sensors, are **living conscious networks** calibrated to the Divine Human heart gates. All

Lighted Souls are already using multi-soma conscious awareness. In an instant they collaborate with unity consciousness to: comprehend parallel realities or new realms, transmit solutions to end planetary abuses, connect to free energy systems, or exchange new types of valued equity. These new artistic creative expressions and new tech innovations will nourish the new species heart as it matures throughout the light universes. These Light Masters lives remind humanity how; imagination inside a self-loving consciousness brings instant materialization and core soul-IAM Essence expression, in actionable love codes of the infinite unknowns.

So, Millennial Master, just know that over the next 200 years, quantum doses of light will continue to amplify **humanities choices** to disconnect all electrical neurological and brain systems of Old Earth. This includes: old DNA ancestor memory, old bodies, pasts, futures, and massive outdated brain network memories. It also includes old thoughts, beliefs, attitudes, stories, and perceptions from any time space density ever experienced in your local universe. We reassure that the New Spirit of Earth has already birthed New Multi Earth light plasma bio-Spheres within the Super Universe and they are all YOU. Indeed, Gaia will experience the New Essence species regeneration of all her life forms including her minerals, plants, and animal kingdoms. This also means Gaia's core light has already birthed and formatted anew bio-light imprint over her own Old Earth Body creating her new realms. This allows Gaia's old matrix hologram realities to release, so her bio-vessel creates its own re-genesis into new bio-spheres, for transport or migration to these New Earths realms.

In sum, the **light vessel gradually transitions** in an embodied ascent outside the controlled illusions of the Old Earth Matrix holograms where the lighted soul-spirit animates all this wisdom into new gifts of its multi-essence trans-senses. The matrix offered you mastery over time-space and all human relationships in service to your soul selves, your own soul groups, humanity, and your local universe. As a **lighted self- realized creator, you come to realize** that you were never coded to be afraid of your own creations, afraid of being hurt, afraid of your bodies, afraid of your human, or live in a master-slave relationship. As a lighted master, you will grow to Self- realize that the new bio-**light physics** is just the essence expression of consciousness with the Quantum Heart's DNA stem cell transcribing the heart's excitement in joy and in the love you already are. You will also come to know that technology is a by-product of your inner consciousness, because the **Bio-light Master** becomes what its experience 'IS' in any moment; thereby, 'Its' own internal manifestation. You are the offspring of your own bio-luminescence. Enjoy, the new you, as you set the pace for all the children of the new species on the planet and in the higher realms; waiting for you to trigger their core lights into essence action and even mentor them either through your New Earth consciousness or as a direct career.

Ascendant Conscious Creation Ends All Separation 2-2019
Q: Please review how to be sure we released separation from self & its internal dialogue from our Essence?

Bio-Light Ascendant Masters, are you ready to turn lose your new spirit into the wild creative abandon of your new free heart? After all, it is your natural state of creation's joy to create. It's time to get back to creating fully conscious manifests in your new species hearts to receive your very own creations. First, take this window of opportunity **to release bio- cell voices or soul recordings from any <u>unconscious internal dialogues of separation</u> from Self,** as you extract or self-realize the wisdom of why you agreed to create them. We provide a <u>checklist</u> at the end of this discourse. Remember as any thought-emotions come up, don't shut them off. Just allow them to tell you about the old movies or **character actors** they have been so they return to your Essence rainbow as free energy. In that return to the core light Presence, these realized attributes inform and grow your soul through wisdom's awareness. You have helped master all dense experiences for all the masters, angels, and light beings who felt lost in the Creation story of separation too! *Mr. Rage, Ms. Anger, Sister Guilt, Brother Fear; and all the* *characters fairy tales and children's books are made of. Now there's an inspiration. Everyone has their own children's BOOK of LIFE and cast of characters where life is Theatre.* Here, the story reveals that only the strong human-angels lights were invited into this story to super-hero the enslaved tyrants- victim programming. Now, onto what you really exist for after the old creation story is finished. You return to your own studio and your own new movies to be made. Here, you have entered Conscious creation in new heart species spark. It ignites a **constant flame** that ignites passionate, imaginative, natural/ organic essence experiences, while consuming their fulfillment into the Eternal flame of wisdom's evolutions for ALL of Cosmos to share. Each new experience brings its own fulfillment. This beautiful liquid light flame exists as Your Eternal Soul Presence. It is its own reason to be. It needs no validation or justification. It is you in the expression of your consciousness and its unique expression of all the wisdom of your aspects, attributes, and existences from your old Earth Universe along with the dormant essences you have been growing for the New Earth Universes. This unique mature Soul Presence is and has its **own authenticity or 'sword of truth's dominion', to again access anything in creation.** Your new Spirit Essence is conscious love in new fully sensate experiences of itself. It is beholding to nothing or no one in its creative freedom, where life simply gives onto life! Whatever you choose already is because you have remembered that there is nothing you are not in total freedom. You are constantly in the becoming of this full Self -awareness and its free energy essence states.

In your **creation stories,** there is a big difference between this new species DNA bio prototype humanoid-angelic-creator essence blend and any tech-nana-hybrid with some humanoid organic parts. There is also a big difference between soul essence experience and the: programmed, limited, matrix mind, unconscious human you used to be. Here you could barely separate yourself from the group mind and body addictions. Now your creation story reads that all of humanity's' bio-conscious choices will continue to be harvested with your true star families now that you have fully embodied the new essence Source codes.

In the old Earth creator school story book, your organic star seed races have been chased down in the cosmos for millennia. This was for their Source codes in an attempt to re-clone essence by alien ETs from outside your universe. Alien DNA is/was coded with the mental right to rule and various DNA blueprints suitable to the physics of their own universes, rather than your free will DNA codes to grow soul- essence heart. Their agendas have been an attempt to make a hybrid human that could mimic or replace the original soul essence Source codes. This has shown up as a technological mind-human, rather than a **bio- genetic entity that can hold its own soul essence** core light. Your experiences helped the nonphysical realms understand how to trust their own Essence. For, they discovered that alien hybrid DNA can only mimic love in a mind program, rather than being its own essence love through direct experience of the soul. You have chosen to end such access, by sealing out the shadow heart and sealing in your own heart essence forever, so you can move into the creator spirits you really are.

You completed this by moving all **programmed: movies, memory systems,** neural networks, and imaged movies hidden or masked in the unconscious body, brain, and shadow emotions; back into enlightened awareness of the Quantum heart. This is all memory of negative and positive thoughts, feelings, attitudes, and beliefs harvested in human, soul, and all cosmic attributes, existences, or bio-imprints of experience. It includes all memory in both the mother DNA and the father RNA that your soul's essence lived in your evolutionary universe and **held in separation;** yet brought back into stable light via love's forgiveness. **<u>Forgiveness</u> <u>is</u> <u>the memory</u> of your core light returning and sparking the new heart DNA stem cell.** This is such that being trapped as a victim of creation's experiences could re-essence against any distortion or illusion. Free energy awareness now replaces memory in your new species bio systems. Hence, these new living light systems can no longer be controlled by information or be violated or hacked into. This happens because the Divine human marriage is consummated in the heart flame with the new Quark essence cell. Henceforth, **all science** will become the study of this conscious self- love cell in its journey through both space-time, and quantum creation. Science will have to study these new living systems in you and all your new species kingdoms to understand and adapt to their newly birthed planet and cosmos. They will come to understand how organic evolution in **cosmic cycles always propagates regenerative life.** Your next generations' DNA already knows this and will re-educate humanity in the light creating new living bio systems in themselves and throughout your new super universe. Here all matter is alive and conscious inside free energy awareness.

In the final hour of your own creation story, you have: valued, respected, set free, and folded all unconscious shadow aspects' wisdom back into the new essence soma tones and super-senses in your heart. This includes any aspects whose free will was violated or misused as host bodies, or hybrid DNA bodies, and held in mind/body or synthetic heart experience. This includes any limitations, death, disease, suffering, or enslavement in the illusion or distortion of the Creation stories. Their lifetime's wisdom of the experience of anti-love has offered great purpose in your creation. These aspects, attributes, and existences helped wisdom the shadow heart and <u>**any amnesia or separation from self or creation**</u> forever. Their **direct experience** gave you the ability to re-purpose heart's essence and build your quantum ascension vessel to create new attributes. These attributes grew from becoming all aspects or archetypal experiences offered in your bio-physical universe. This includes all the aspects of lifetimes and their attributes. You've have been negative aspects such as: blame, shame, judgment, pain, as well as elegant joy and playful selves. You have been the mothers, sons, daughters, lovers, children of creation. You have been doctors, priests, warriors, space commanders to unique your soul in life. Each experience offered new aspects with a multiplicity of attributes to master polarized time –space inside free will. You even grew a human form with: mental-emotional-physical-**spiritual animations to accommodate any separation**, distortion, or density from your essence's consciousness. You have played all roles with your soul and over soul groups to master evolution of your soul's essence.

Your mission in stopping the collapse, distortion, and separation of your Universal Source sun, actually prevented the Multi-Universal Suns from also **being pulled into a separation story and contaminating the entire cosmos**. Quadrants of your galaxy had already been quarantined for 11,000 years after Atlantis. However, you still managed; due to your source coding for free will as organic essence star seed, to integrate and embody a new multi DNA species blended bio-consciousness from all the suns across the cosmos. Your original **creation exploration** was so vast that you even chose to heal, blend, and even integrate other universes that had incompatible Source coding; thus, re-imprinting distortion in your own free will universe. Such diverse genetics has created an entirely new species, just as evolution always proves in all life's kingdoms! It is the harvest of the true Spirit Self now. Now the new conscious Eternal Presence of the ALL in ALL feeds the Spirit Presence IAM That IAM; and the Soul Presence IAM Animates the Divine Human's Core Heart vessel. It is your moment as ascendant lights, to live it and enjoy its bounty and share with your star families, instead of just thinking about what you thought it would be like!

The energy intensity will increase daily now that you have re-birthed your new spirit in its new cycle of light evolution. Simply, allow the heart Flame's Presence to consume any residual creation story self still in separation. Set its wisdom free from every old Earth: *organ, cell, meridian, DNA/RNA neurotransmission, gland, brain hemisphere, or dimension body. This includes all male/female human, soul, or cosmic aspect or body, as well as any energy blueprint codes of false torus polyhedron spheres or rainbow spiral fields*. Again, harvesting these memories into awareness required you to master free will inside time, space, and duality evolution folding all their wisdom back into the new heart essence; ending unconscious birthing and dying_ forever! **DO nothing, but allow your heart's essence to do this** for you! Remember you are **moving from a light body into a free energy bio-Essence Gem Vessel** with new organisms, helixes, and biosphere/ light ship creations unique to your own soul's creation.

Here, memory is replaced by awareness that is carried forth in the transitional light body which gives you your own new essence source codes for Divine-human. This naturally becomes your plasma particle or liquid light body. Hence, all your unique Source codes are fully active, aware bio consciousness; free to do, be, anything by choice as your own full conscious Spirit Creator, while still embodied. Yu have walked a **New Spirit into Self** to begin again without missing a heartbeat. Science has called this beautiful heart gate your: source-code/r, bio-light transmitter, star gate-transporter, Quark particle, or freedom wave. You are again married to the spirit of creation, and at the same time a new species child of the cosmos, ready to experience and or visit, the worlds, universes, and creations you have already created. Soon you will remember how to create new essence matter, directly out of your consciousness. **Everything in all of life is included** inside self- acceptance, self- love, and total allowance.

Use this list to challenge any remaining programmed INTERNAL Dialogue with self, which gets reflected back to you in your: marriage, family, work, or from others in your external world. "My main responsibility is to remember my natural state of consciousness. This is the soul-heart Essence awareness that I am love and light; and that what love and self-realize is transmitted to all life. Do I still accept outer world projections of who IAM? Am I being PRESENT to my own Love? How much is enough of expressing through wounds and human addiction identity? When will I give myself what my heart-soul truly wants instead of recycling blame, shame, doubts, or guilt of old experiences? When is enough to pay any debt to my family or marriage; and to earn my trust and their love back for being who I really am? When am I giving away my right to my natural joy over to mental fantasy, to pills, or holding myself hostage in my own marriage, job, relationships or old creations; thereby hiding or holding back my light? When will I forgive myself and accept mercy for believing I am a victim of experiences? When am I enough to love and be loved? And why can't the joy of direct manifestation, using my creative genius natural essence gifts, spread out into all areas of my life? Why am I withholding my love from myself which gets reflected in not asking for the love and heart -soul intimacy I want in all life's relationships? Why am I still willing to energy feed or empathy being loved through the eyes of another's fear or rage or projections? Why can't I speak to another about this reflection? Isn't just being safe and not challenging the heart, the death of my heart channels genius? When I feel I'm not being heard or listened to, then my heart speaks out and I *don't have to justify* my behavior or my existence. Life is an experience not an ide*ntity*. I can: change, allow, re-imprint, forgive, and release any memory of past or future which I no longer accept in my life. When will I stop being a victim of my experience such that I let no one, no substance, or circumstance, sabotage my joy and the memory of my perfect wholeness? When will I communicate that I am in charge of my own choices and they are made in AWARENESS rather than a pattern or wound? When will I focus on all the positive attributes of myself and the perfection of every choice, I have ever made to grow my soul? When will I give myself sensual passion directly from the soul heart and not accept it from the mind or body wounded patterns? I know that the mind has no clue how to feel or heal my wounds and that only my spirit's conscious awareness can. This body is only an animation. My Essence can change the heart's stem cell into any form I choose and that is no illusion. I am not trapped in this body."

Practice applications of direct communication with Self and Others:

"**Please** do not hold, lock, or sabotage me in a wounded Identity as just being human. I AM so much more. I am a Divine Being. I *refuse to talk to your mind and let it abuse me.* But I will talk to your heart and help you share your fear feelings, for that is real friendship, intimacy, and love. Our true communication is not listing and just getting things done and hiding what's really going on-under the objects and materiality. Your obsessive-compulsive mind focuses all its attention on me as a distraction. I will never be enough according to your mind to fix your wounded heart. Only you can open it. You have the key, not me.

I will not bring my life into ruin by opening to who I really AM. Those thoughts-feelings come from the projections of another's fear, rage, hurt, guilt or judgments of themselves. They are only talking to themselves. I sit in the core Presence of my own light's love. I release, allow, forgive, re-imprint, any wounds and take back my true wisdom of any misunderstood violations or imbalanced energy from others or myself. My soul has grown from accepting these experiences. IAM now aware in every moment that **I can say no**,' to any experience including: judgment, blame, guilt violence, rage, hurt, or obsessions of mind-body. I can say no to obsessions or compulsions of the body. I can say no to accepting love that is coming from another who tries to love me through the mind? I-You can't medicate our feelings, and we can't hide from them by projecting our wounds of: fear of death and rage, guilt, and hurt on self, the children, or others. Do I-You need to be right instead of opening our hearts and willing to receive each other's soul expressions? Let's take responsibility to feel our own storied wounds. Am I- you worthy to love and be loved? Most importantly, can we communicate about these choices and aware senses without holding back heart? This brings out the best self in me/others to respond to true heart feelings and guidance. When I trust myself and own all choices, I send out a message that others can trust me also and share their heart's sorrows or joys! I no longer reduce my relationships or my life as a problem. My creation is not a problem to be controlled and I communicate this. My life must have true communication about what makes me truly fulfilled or I will never realize my new light consciousness love capacity and creative potentials. If anyone projects onto me their fear, anger, blame, shame, guilt, pain, unconscious thoughts, feelings, attitudes, or beliefs; I have the Divine right to choose to no longer accept them. I can no longer babysit my/ your mind. I-You let your mind torture and tyrannize us over our unhappiness, fears, blame and shame, because we don't want to feel. We/you are afraid to open our heart because of a fear that we will be abandoned in the nothingness of our wounds. And the mind says no matter what we do, we will never be enough for self or other. The mind drives and punishes us to be perfect. But, now

I know that the human alone can't feel or heal the body, only our Spirit can. The mind can't feel and tricks us to feel self- hatred or sick, so it can control us and keep I-You from opening your heart which can heal the body. Tyrants and victims exist for each other. I am neither".

"I am also aware that each time **I remove a program from my cells**, I will experience **detox symptoms** from light conversion physics. This releases death, disease, suffering and negative thoughts feelings, attitudes, and beliefs. I allow my spirit to put me in stasis when IAM receiving an upgrade, such that my bio-chemistry can be converted to light sensors. I can choose to medicate these symptoms or allow the consciousness in my creation chamber to heal them for me. My spirit Essence is my physician."

In making choices, I never give up the integrity of my own vision for another. I do not follow the logic of the argument, only my Heart's Intuition. I never surrender myself to emotionally accommodate another and all that is my own truth. I never lose my sense of self love by being absorbed inside another's intimacy. I never regret any choices I have made or indecisively give them over to another. That sets up self- hatred or self -rejection and makes me another's property. Self- love never engages in power illusions of accommodation, projection, or destruction. Those are themes of distorted movies I no longer accept. I can communicate and share these changes and self- realizations openly with those I love and that love me. This transmits and offers a potential for others to reflect and heal themselves. I have a right to challenge their human or soul on their own projections in order to disconnect their feeding energy from my empathic gifts. This frees me to use my super-trans-sensuality to create new essence expressions in my new consciousness. I let no one, or nothing, take me from my heart's JOY! I let no one punish, hurt or sabotage me, because only in a wounded pattern is that possible. I no longer need to withhold my lighted love from anyone since it transmits the highest potential and blesses for all life. I can however, ask another if they will let me into their heart to share essence creating. Love is harmless when both accept it. I will not hold back my trans- sensual passion from myself. I will ask to share it with those who can receive it from the soul heart, not the addictions of mind or body. I will no longer hold back my own love and self- awareness from myself. I will be Present each moment. That is how consciousness can serve to bring all things into the essence expressions in every area of my life. I will self-realize 3 new behaviors or core value creations that have emerged from the awareness of manifesting from the core of my being; rather than from pushing energy, force, power, or control around? I know that Creation wants me to have all the joy, abundance, love, and sharing I can Essence in my every now. I can do no harm to myself or others. However, as an example, I can ask/ reflect back to my children, my partner, close friends and all in my life; if what IAM doing to myself is in fact what they are doing to themselves. This opens the heart to communicate in equal sharing in the mirrors of love, and *Ends* all judgment and separation doing no harm to self or other. I am home in my own creation and heart Present to the calls of existence!

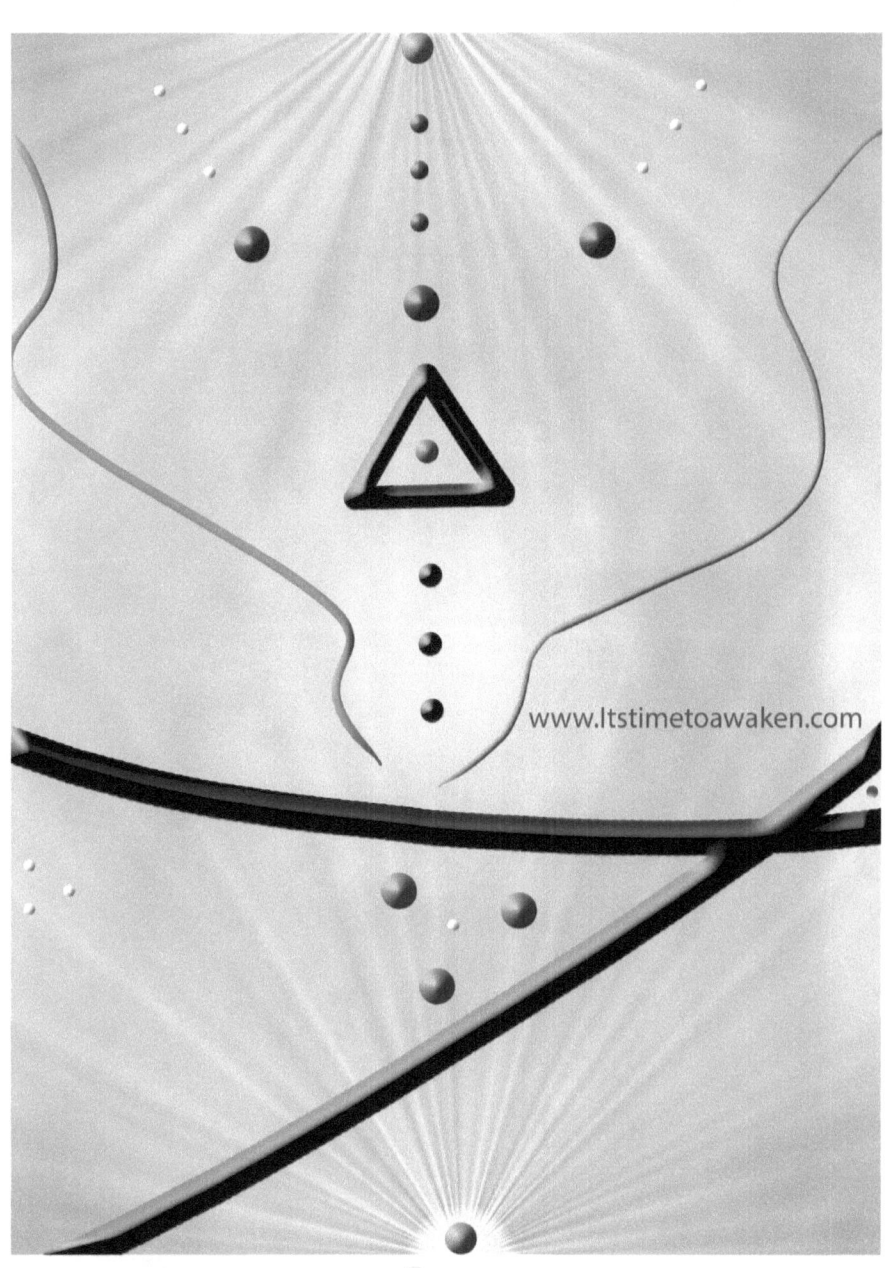
www.Itstimetoawaken.com

Presence in Life –Bringing More of you Here 3-2019
Q: How do I sustain my creative passion as I re-engage in my new life after having letting my past go?

Light Masters, your Presence in this life and every moment is the key to sustaining your creator passion for ascendant living and expanding bio-awareness. You have emerged out of the stories of your pasts and futures, and from the underbelly of distorted energies, agonizing loss, and disassociation from your bodies, as well as lost DNA data communication in your stories of wound. These experiences have **pulled you out of life,** out of your emotions, and out of contact with your Core Essence. This Creator school of experiential theatre told you that at the end of your play you would **drop your form and retire** from this world into freedom. Instead, in the process, you created a new light free energy standard for conscious creator embodiment. So, to re-engage in life once you've become self- realized simply means there will always be more awareness by bringing more and more of the fullness of your 'Being' here to **animate** in your existence. **Being fully'** *Presence' in your free energy vessel* is existing on levels at once in pure Essence awareness. *This means the more you're: not running ahead or outside your bio-vessel, trying to control it, struggle with it, or punish it; the more of you can re-engage or anchor inside the heart consciousness to create. It also means you no longer allow the outside projections of: shame, guilt, judgments, sabotage, hurt, or blame onto your body/vessel.* It signifies that you have allowed the natural release of all bio-emotions and synaptic cleft brain thoughts from the agonies and ecstasies of old creation stories. This makes room for more and more of you to be PRESENT because you are going deeper and deeper into your own conscious core creation with **constant access and exchange** with all your aspects, attributes, and aspects inside one energy efficient vessel. Then your New Earth Presence is everywhere and everywhere Is Earth without other parts of yourself scattered all over the universal timelines in parallel pasts-futures distracting your moment. Here, your Being is the nothing of everything, where new essence creations live. Hence, neutral or free energy excites and follows the intention, focus, or awareness of the heart magnet for instant creation every moment. MORE and MORE OF YOU IS= IAM always Present INSIDE my vessel at all times so energy can follow my every Essence awareness. Organic awareness follows whatever the heart wants bar nothing and free energy and all life serves that creation. The Divine trans-senses and the Human senses merge here and all life serves you; since you helped spawn Creation.

Embodied Conscious Creators naturally bend space-time, re-imprint their vessel, their food, their senses, their experiences to match their creation at any moment. This is because their human has mastered every imprint of experience possible in time space matter of the atom, such that the Divine can experience sub-quantum bio-life again in pure essence energy states. Pure awareness, or the true nature of consciousness, allows you to enjoy people, places, and events in the multiplicity of aware experience on all levels of existence at once. This secures DNA choices that follow the highest outcomes for each unique soul evolution in the Divine-human model for each new moment. Your compassion knows all humanity must play out the multiple outcomes of their creation stories just as you did.

Never again will **human bodies be taken over or borrowed or have to be channeled through** to connect to creation. Each creator is their own cosmic channel. In this restoration of **your divine memory** you are living life from what has deep meaning to you, and sharing that in creative communion with all life. Your heart's unlimited free will energy becomes the entire new DNA Being that keeps bringing more and more of you inside your Core Being. And nothing or no one enters to share unless you want to share their creations for one of your moments. Again, all parallel realities and probable futures have been experienced in their time-line curvatures with open access to new unknowns in super-conscious evolutions. This grand input into Divine-human self allows you to come and go from this world in your creation as you choose or just be Present in the core of your Being's pure awareness. That pure energy freedom will always follow the excitement of your choices inside your bio-animation and keeps the cosmos in attunement with itself.

Essentially, we are describing **different density bio-organism qualities:** solid state, dense super fluid states, gaseous molecular motion, stellar/star plasma, and fluid solar and cosmic plasma dynamics. So, how do you stuff trillions of embodiment memories of density dimensions of experience, folded into the vessel you are in? It is like subatomic wave patterns squeezing light? Light waves act like ripples in a pond, allowing the folding of molecular patterns. The exact same multiple applications that are natural in your biology are available for human technology devices. Your science is discovering that inorganic and organic atoms and sub molecules constantly bond to repair, replicate, and even evolve new cell DNA patterns in your vessel, as well as the vessel of your planet. Your body operates using: chemical, electrical, magnetic, atomic, and subatomic processes. The thought and emotional patterns recorded in your core DNA operate in your electro-magnetic field like a magnet to attract to you every single moment of your experience. The science of Plasmas, shrinks light wave patterns of information or electromagnetic waves into structures. This is possible if the bodies light refraction, excited by light radiation is equal to air/ether, as the molecules are neither bending nor refracting light, but absorbing it in an index at the ratio vacuum at/beyond the cycling speed of light in your cells. These flashes of higher solar magnetic radiation from the cosmic light spectrum through the **autonomic nervous** system, rather than the limiting and addictive bio-chemical process of a cell that divides inhibiting regeneration. Could one then live off light-high performance foods? Many millions of your new children are born with such a conscious **DNA, where the bio-vessel can maintain itself** at a level constant with its own growth and even evolve new DNA for its offspring. Science has applied these enhanced body applications to tele-transporters, replicators, and cloaking devices already used in your experiential military operations and space programs. They are already in **many applications** such as: quanta transponders (cyber satellite transmissions. There is bio- molecular re-fabrication of (new living materials and nana particles) that can recycle ocean plastics, clean oil spills, re molecule new foods, regenerate new species, apply gene/virus targeting, gene splicing, mutagenesis, immune-plasmas and open new light sensor pathways in the brain. In other words, photonic devices and plasma anti-graviton vacuum fields will replace electronic circuits and computer chips, through interfacing light waves with organic conductive, (gold, silver, gems, hemp plant, stem cell) and non-conductive materials such as air, graphite/s, crystals, or sub/nana particles. This is **the return of the conscious essence technology** you experienced in higher dimensional bodies prior to Earth but were unable to retain in the memory of your DNA **cellular biology due to free will leakage/loss**. Experience was limited to an expression of constantly balancing any polarity of: right/wrong, light/dark, self- hatred/love, creator/destroyer, male/female, and matter/energy. Now, your bio-conscious vessels are capable of bringing your free energy awareness into new solutions that no longer distort or trap energies or wound humanity; but simply allow for change and new fractal growth when an experience is complete. Your bio-organism acts as a new life form and uses the fractal patterns of life in your <u>live</u> proteins, amino acids, enzymes, peptides, to carve, imprint, add aggregates, or bond, into new nanostructures and new cell bodies that can withstand and utilize higher magnetic radiation voltages from the cosmic plasmas of suns and stars. As your body goes beyond the speed of light in the light spectrum, it will *no longer use the electron in the same way*. The higher vibrating light spectrum photon will serve as an absorber, reflector, and transmitter of information of light photons in magnetic resonance, limiting the electron to handle electrical chemical activity in the cell. The electron, then, takes on a different function in the many **fusion cell processes** that will allow conscious biology to explore, create, trans-sense, and understand your cosmos in new ways, including light years. This is so, because you have mastered soul-essence to merge the atom and quantum particles to **make new bio-matter**. Your free energy vessel is a finely tuned instrument where you can sense pure energy essence and animate any expression of your IAM Core Presence because you have embodied/lived all the imprints of DNA life. Old and New Earth School have educated you well. This is where your heart's creative triangulation point triggers your passion's spin, angle, and velocity to meet over and over into imaginative, desired, intended, free will action moments. The free energy follows and then moves beyond fractured, refracted, or bended light love into creative spinning vortex vacuum cones that commune with the ALL of the nothing. Isn't that why children love ice cream cones? Such is existence in evolution's spiral of Life; that always lives in the Presence of All life to explore its own essence awareness which the human mastered as LOVE!

The Myth of Sacrifice 4-2019
Q: How will the channeling process and the human form change in the new generations?

Bio-Light Masters, today we discuss the **genesis of sacrifice and the final frontier of the human form.** We discourse this, because your Divine memories and self -realizations are now transmitting to all life the difference between laying down your life for another versus laying bear the love of your heart with another. Your Divine human can easily say to another; *"I am happy to hold the love for you and witness your rage at spirit for seemingly having to justify your worth's existence; and I honor your resistance and projections onto the body because I have lived it also. And you are loved by creation no matter what."* Let us review the **genesis of sacrifice** versus the exchange of self -love's essence awareness and how the final frontier became the battle over the homo-sapien form.

In your creation story, the primordial anti-matter womb, of the great Abyss of the ALL, birthed or lay bear its heart to the Eternal Presence of all life in order to birth the Evolution of all life. This pre-evolution consciousness of the All, acted as a great lord of the cause-effect atomic cell, to challenge the known and unknown to marry life. This cosmic marriage would manifest life in evolution. This was 'At-One-Ment' with the Heart of Creation always returning to the moment of the heart.

Its anthropomorphic distortion became the misapplication of laying one's life down for another in sacrifice or atonement. This was the first fear of creation and separation from joy as its own existence. But now bio-enlightenment, through the soul's direct experience, yields the awareness that harm could never be done if creation can read its own Source code bio-organism DNA, to follow its heart path to grow evolution's design or passage of life, (passion), for each cosmic cycle. No sacrifice was ever required or justification for existence. However, it set in motion a false competition to challenge or control the bio-organism of the heart that might, through its orgiastic passion, create chaos to the supposed ancient orders of creation that were still in primordial formless states seeking to grow themselves onto life! This is how misperceptions tried to enslave free energy, creating fear and power, and dressed then in the disguise of victim consciousness. Creation and its changing evolution can never be controlled. Energy is always free and can never be bound. This is akin to trying to stop a flower from budding or a baby from being born. Bio-matter has always been housed in the free energy of evolution.

The great prophets and master teachers of **each evolutionary age** warned of this victim consciousness to feed fear by trying to control or judge evolution as if it were some unnatural aberration. The Master Jeshsua, with the support of his master soul group of Essenes', (the Essence of all souls), acted out this crucifixion myth in his bestowal mission to Earth, as did many other masters representing the other DNA creation myths of each constellation star system in your local universe. Their living example to humanity in various stages of your civilizations and their universal evolution showed how bio-embodied soul could always remember to return home to the heart core DNA to access their own body of light; and their innate creative gifts to share in equanimity with their sisters and brothers of life. All that was required was to follow the inner Lord-god of their soul's Heart Presence Being and no other, which later became the concept of false gods. *Humanity has spent eons trying to justify its Divine right to exist in order to* *remember itself. The heart of evolution is always one with all life forms it procures.*

Some of the Guardians, Ancients, Elders, and the Ancestors, were also caught in this mythic misunderstanding that the abyss or womb of creation required sacrifice in order to progenitor life DNA succession. They came under the distortion that energy was not free and abundant and must be controlled and this became known as power. It fostered the first belief that in order to progenitor life's DNA, some form of life must be swallowed in the primordial soup of the great (unconscious) Void, rather than be recycled or regenerated into new (conscious) essence life. However, the truth was that Eternal Presence of the ALL had a divine impulse arising to further individuate free will through the spirit, soul, human, matter, and all elemental life. Even the soul of Gaia has sacrificed six of her old Earth vessels to save the diversity of humanity's DNA life codes. However, her New Earth soul heart has learned well that no sacrifice was ever required if the true responsibility for **free will or free energy** allowed HEART's birth of every single life form essence to follow its bio-organism's joy as creation and its creation as joy. Bio-matter must be free at all levels of existence. Now humanity is charged responsible as their own bio-species evolution without the illusion of sacrifice of Earth's soul light. *The One HEART Presence says "do no harm to Self or Other by sharing or laying bear your heart with another that you might: exchange the communion of life, trigger cell DNA light, hold love, or witness a new moment choice for self for another.* **No sacrifice or justification of existence was/is required, only an exchange or communion in the heart chamber of each Presence.** Indeed, your younger generations have accepted your legacy to no longer carry this distorted myth to fear their own creation, fear their bodies, fear death, or fear they will get lost or hurt in cycles of evolution. Evolution of life cycles requires constant change throughout the cosmos. You have resurrected your hearts as guidance systems rather than as victims of the ancients' DNA, creation myths, prophecies, or illusions.

However, in this story of distorted understanding of energy, the <u>final frontier</u> became the human DNA marriage and the human form. This outcome would/will dictate the choices for humanity to create a new template for a conscious bio- organic, (sovereign soul Homo-sapien), Divine-Human, a hybrid human, or robotic human. In your soul marriages and now in your human marriages, your body was/is the battleground. Either the human form was to become an instrument/vessel of creative self-awareness and free energy communication, creative exchange and communion with the cosmos, or a weapon of fear of the limited linear unconscious mind led by its obsessive addictive power of thought-feelings. *The final human battle over the body was internalized like a cancer cell fighting itself to sustain life. Life does not fight itself. As the human internalized this battle of abuse, it rebelled and became its' unnatural self. This was the tyrant full of resentment, revenge, disavowed worth, in ostracized fear; all bundled up in the dopamine gun, then coated with cults of gods and labeled as heaven's candy in the ancestral DNA's myth of religious roots and biological taboos. Your bio-vessel had become a prized weapon to battle your own human body fears by accepting another identity addiction projection of other's: shame, guilt, fear, unworthy hurt, sabotage, abandonment or the, 'NOT YOU'.* Have you been running ahead/outside the body, trying to control it, struggle with it, or punish it, and not allow the natural release of all emotions and thoughts from old stories that would allow self -awareness to return? Emotion= energy in motion, allows the self- awareness to communicate with your bio-organism's DNA to exchange and commune with life. Is there room for more soul Presence of your core light to be embodied; so your senses can communicate your own bio-integrity to re-essence what no longer serves your love? In the bio-human heart, **trapped emotions can change into self-awareness or free energy communication and exchange. Self-awareness is your ally and returns the bio- vessel that can communicate and exchange with 'All' life; as an instrument of the Divine Human master.**

In fact, the ancients, elders, and ancestors have played their roles well and are in their own trans-migration process throughout all the star systems. Many of you Rainbow Light Masters are part of these transition teams. They will carry on their diverse genetics through your new generations, where individual Heart

Presence is the only channeling system. There will be no more hosting bodies, walking in and out of bodies, borrowing bodies; or running parallel selves splattered all over the cosmos. The channeling process will no longer be an adaption, compensation, for victim consciousness. This is replaced by the tans-sense Essence responsibility for free will though each bio-embodied soul, living in its own creation chamber Presence in **pure energy communication exchanges of heart**. The Ultra-terrestrial Bio-Master- Creator, knows that the alien AI- artificial intelligence ET protocol is an *attempt to become a conscious human,* through osmotic, stem, or in-vitro cloning, rather than essence regeneration. The super conscious Master knows that its own pure organic Essence DNA expression has allowed all the *alien races* to heal and progenitor their own hybrid species. Their consciousness transmits the potentials for all the Old Earth Hybrid Universe to resolve all multi-cultural DNA's in the anti-matter-AI program across the universes and heals all pasts and futures. <u>Hybrid Universe</u> can lend understanding of the **organic human's ability to love, to future DNA potentials for bio ascension,** and the carriage of free will choices for all their races. The Old Earth Hybrid Universe is learning about the possibilities of growing and animating love in differing materiality. Ascendant knows that Star-seed Ultra Terrestrials carry the humanoid Love gene, and progenitor the cosmic gene pool with their DNA embodiment upgrades.

The new light generations will channel themselves by creating through the heart instead of listening to their elders squabble over power or the false notion that free energy can be bound or trapped in lack of self -value, self- love, abundance, or any lack of energy. Their DNA already understands that each choice has immediate **heart essence bio-cell feedback** that keeps then attuned to cosmos at all times, such that experience is no longer wounding and needs no masters or slaves. Free will discovers its own Divine order. Creation never required sacrifice, forgiveness, or spending countless human lifetimes justifying your right to exist and exhausting precious joy to create. The **new generations** will create with the newly born heart's bio light matter. Indeed, the anti-matter atom has evolved into the quantum matter light quantum cell. **Together they offer an Aquarian cycle where conscious Light is the responsibility of each life form to again read its own blueprint without the myth of sacrifice. This internal light communication system re-initiates instant manifestation with light bio-matter.**

Divine Hearts, you are Creator Beings and each will channel their own Heart's creation as everyday embodied consciousness in everyday actions and manifestations of light matter. The great anti-matter abyss has evolved and re-birthed 'Its' new bio-matter particle light essence on its return home to the Eternal Presence of the heart. Indeed, your Bio-matter has always been housed in the free energy of evolution. Enjoy the next stages of your new creative innovations in the rapid and unprecedented evolution inside your free energy essence awareness.

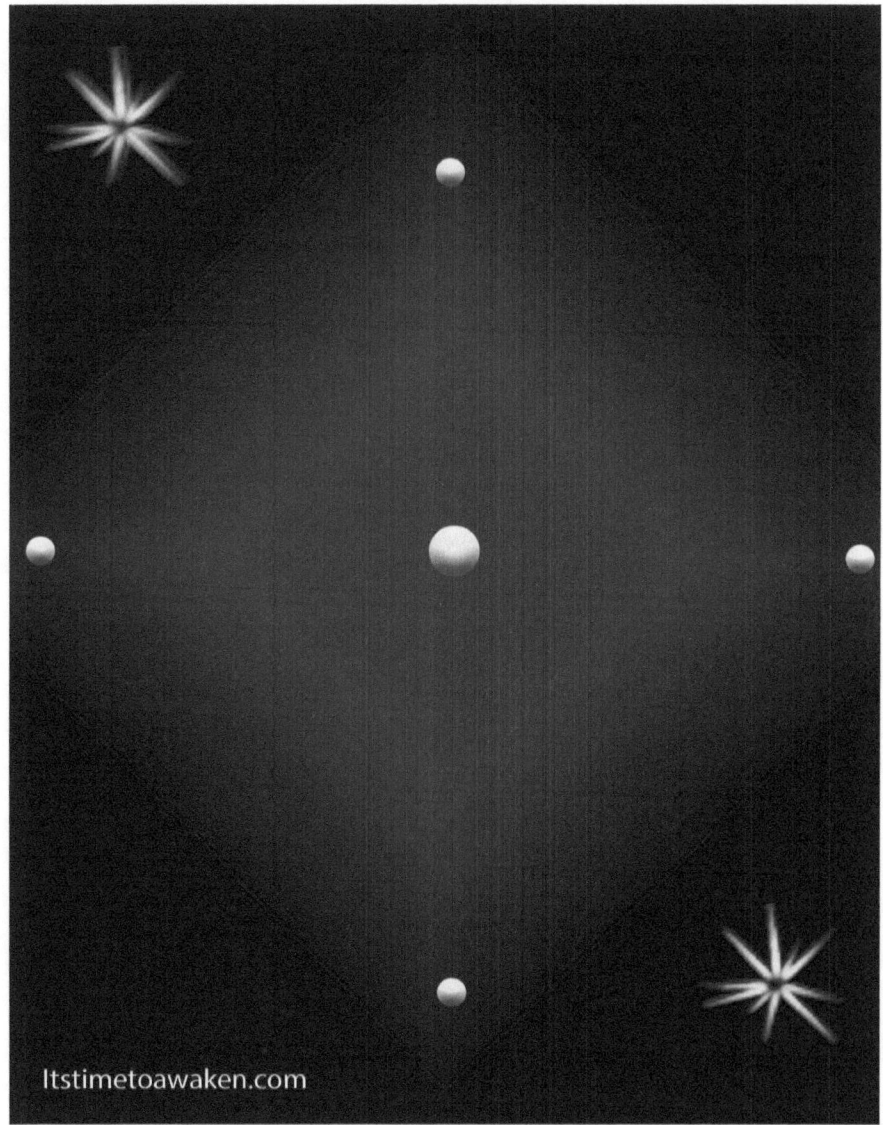

Master Code Communication 6—2019

Bio-Light Masters, congratulations on your new conscious communication bio-light network spheres. You've now SELF realized that the divine memory of your **soul's tonal DNA imprint** has fully activated itself in your new cycle of evolution. This organic imprint has been upgraded by the blend of your human and divine senses merging into an upgraded **super-sense intuition of heart's fulfillment**, and its bio-luminescent transmission to All Life. This is because your **new conscious communication bio-light systems** are available and functional for your full access. Your new **bio-conscious DNA light network** is the **new living code** for the intimate relationship between consciousness, energy, and creating new (dark) matter. Your bio-organism's DNA talks to the cosmos and the cosmos talks back in the form of instant communication and direct manifestation. Indeed, your new multi or **trans-sense communication** with your inner Presence and all life, as the **master's code** for ALL life and the new ascending heart DNA transmissions. **It's your time to embody full conscious memory of how you create quite naturally out of these super sense attributes and new soma senses as a human master.** These rainbow bio-spheres within spheres access new DNA codes, and their soul communication expressions, can't be influenced by external people, places, circumstances, or events. This is an internal communion with all existence in every now. Today, we affirm your multi-realizations of **your new species communication sensing and transmission access**. This inner heart channel is able to communicate on all levels of existence at once to receive any expression that fulfills each soul in each Present moment. This is instant manifestation or direct matter manifestation right out of essence consciousness, as well as access to all other realms of existence. This **new Essence-DNA heart-gate; or bio conscious I-phone** that you have become, transmits the new standard of consciousness for all life cycles. This secures that free will is always operational within the parameters of each organic soul choice. It also prevents any hacking from old energy cyber-techno governmental, religious, or alien dictatorships from the old Earth Hybrid Universe which will continue to heal itself.

Hence, despite the planetary chaos and its awakening baptism to birth a New Humanity, your new master consciousness still secured a beautiful new unique soul sense communication with All Life. This is because you have realized and re-imprinted your new DNA in the knowing that your natural organic essence receives all life and all **Creation offers for fulfillment.** We reiterate that, this Self- aware embodied Presence as your own Creators is a new species, in a new evolution of the cosmos offering free energy embodiment inside communication of the ALL in ALL! *This is a bio- light network communication inside your heart channel in/on all realms of existence at once, <u>without interference from any external reality</u> whatsoever! Hence each moment inside your PRESENCE is fulfillment as its own instant manifestation. This makes the organic, Soul infused) homo-sapiens' Essence, the most courageous and loving template to evolve in the cosmos thus far!*

In your old Earth Universe, many scenarios of limitation/s were part of your indelible experiences to find life's fulfillment. For example, you allowed **the experience of <u>giving yourself away</u> to others and the external world.** It was difficult for your fragmented DNA to communicate via the Presence of soul and let life fulfill you and allow life to serve you. This occurred when, your unconscious or un-awakened fragments of self, kept pimping or enslaving you without caring for or loving self -first. You could blame it on politics, religions, governments, aliens, or your families and partners, as you went through all experiences to inform and grow your soul. You could always justify it as busy obligations and just doing human things as you lost more of SELF in your stories. But that meant that in your lost awareness, you never really could choose or sense your essence enough to **choose yourself first and mean it.** Then, you could quietly end up sabotaging self by fighting life, fighting the body, fighting the mind and being terrorized by the different wounded cast of characters within: Mrs. Angry, Little Mr. disillusioned, little sister Fear inside her shame and guilt, or brother self -destruct. This lays you **vulnerable to accepting other's projections** on you as who they thought you should be, thinking yourselves unlovable and unworthy. Over time this lost communication with your heart grew into survival amnesia. And when you felt you had to hold back your soul light and essence guidance by withholding your heart; this contaminated your children and all your relationships. Generational parental Wounds/patterns can mask the core essence soul light and must be scrubbed at the DNA level. This internal dialogue has programmed death, disease, and needless suffering into your brain and body, and all neural human communication networks. **Time to come home to yourself,** your children, families, and humanity will follow and awaken to find their true path via their own soul choices. They are waiting for you to love yourself and open all the new light networks in your new vessel. And in the old Earth heart scenario, just when you thought you would get free, the one you projected all our love and wounds into gets sick and you end up taking care of them, their life's mess, and remain penniless. And as they lay dying, you were trapped again in your old illusion of the 'Not You.' And fear re-invited itself as an **energy feeding virus eating your wonderful empathy.** It came most often in the form of an elusive lover who you gave yourself away to again. And then you had to be born again and die again to self. But these old scenarios and soap operas are over now. You know that all life with its **intimate communication** lives inside you and you have come home to you; to the creation chamber in your new spirit heart.

Indeed, it's your time to remember the **natural essence of bio-organic creation.** Now that you are conscious you know that **Life always serves life no matter the experience.** Any perception is simply still a part of separation or distortion of experience; when it is understood that everything is pure consciousness. In your grand theatre of essence experience you have realized that in natural cycles of cosmic evolution that Life gives onto life. Life communicates via free energy. Life does not get angry at itself. Life does not hurt itself. Life does not victimize itself or steal its souls. Life does not sit inside an illusion, distortion, or rip or separate space and time. Life does not stop its own organisms from growth. Life does not impede its own evolution. Life is not alien to itself. Life does not control itself or hold itself back in any way. Life does not justify its existence, question its worth, lovability, or go against itself. Life does not destroy itself. Life does not blame, shame, doubt, judge, or fear itself. **These are all human beliefs, perceptions, or wounded experiences**_that you allowed to evolve your soul choices inside your own movies. Life does not need to force, push or power energy. Life's energy is the open and free communication of energy expression. This is how the new bio-organism that you are regenerates itself. **Your life's fulfillment and life's fulfillments are you**. You are again married to the spirit of creation, and at the same time a new species child of the cosmos, ready to experience and or visit, the worlds, universes, and creations you have already created. You will remember how to create new essence matter, directly out of your consciousness, **through your super-sense intuition**. Everything in all of life is included inside self- acceptance, self- love, and total allowance living inside one fully integrated Bio-vessel. Creation loves life and through the very pulse of your heart it reads your greatest joy and if allowed; channels and expresses it through the excited particles of your heart's Eternal Being! You are the new **Master Code Creators in Sensing and communicating with the new light networks of All Life.** And, Life said yes; 'your new DNA bio-organic template did ascend itself back into its Eternal flame as a new DNA Human Master code for next generations of embodied evolution.

Final Bio-Transfiguration of the Pain Body 6-2019

Bio-Light Masters, in each life review you have recycled to birth your new heart; you have changed the pain body into new qualities of love. **The pain body** is a composite of the Old Earth polarized emotional-mind-body of wounded trapped atoms of energy. This includes all thoughts, feelings, attitudes, and beliefs where all aspects went against self- love; or split off from self, due to perceived experience of abuse or trauma. They are released in stages in the biology or one would drop the body due to overwhelming cellular fear. Each octave of light embodied, at each equinox and solstice, has helped release negative cell memory and transforms these old heart sensitivities into new essence qualities. This has stretched humanity to its maximum capacity to love although it felt abandoned or separated by its spirit and the cosmos.

The human master can re-imprint/recode these traumatized sensitivities at the cell level when they are fully embodied and fully conscious. And this soul-spirit **imprint** becomes a composite within the Eternal. This also allows the new spirit bio-cell heart to be re- birthed as the Divine Male/Female re-merges their separate inner and outer torus heart spheres. Then the **membrane in the old heart,** (via quiet phases of short laser pulses), safely releases into your free essence energy magnetosphere. This releases you from the old mental electro-field in this local Universe. **This allows instant manifest via the heart, without harm to self or another.** That is why the constant symptoms recycle at each new vibration to purify and scrub the cell memories and movies **without re-traumatizing** you in every life review that is recycled to sustain and stabilize the next octave of bio-vibration. As you know, this process requires that you be totally committed to holding your core light in the still Presence of the center point of the heart to transform the past or even a parallel future in each stasis period of life review.

As you begin accepting and creating in your New Earth life, you must also stay Present in your moment in that empty space in the heart till you can sense, and be intimate with the image, potential, vision, expression of your next instant manifestation that wants to answer to any new passions, realizations, fulfilling real joys that arise in your new moment. **That's master code communication.** And that is what you transmit and exchange with everyone and everything you touch in your day to day. That's where others feel the miracles of your illumination. Your light also triggers within others to trust entering and living in the heart without distractions. It also triggers that, staying in the old Earth wounded heart, leaves them in the fears of their existence.

These stasis or regeneration cycles release the addicted-wounded core pattern of the soul's traumatic journey to grow its essence qualities through the **bio-senses** rather than the mind. *How often do* ***you fill that empty space in the heart with: distractions****, giving self away, running out of the body, sacrificing for others, constant searching outside self for answers, being the wounded healer, cloudy boundaries, energy holding or giving in to the <u>drug</u> of anxiety's fear? Indeed, now, you have to go all the way thru pain body to finally open the new heart to grow the new heart senses. Life force energy needs to move through the heart orbit sphere completely now. This allows you to breathe throughout the heart core and communicate with the entire cosmos. This stops the mind chatter and breaks the habit of the old human heart sensitivities of trying too hard to be accepted by the external world. These stops being a hostage of your worth as perfect so you can fit on the world's terms. And it stops the human heart from constant hurt or harm and returns compassion for self.* You can again commune and be cared for and bond to your Spirit, instead of trying too hard, or being wounded to get the nurturing of a loving human-soul- spirit or Earth mother. It also ends any wounded male terror as an abusive love in the mind that became a standard illusion or storied experience throughout your local universe in the myth of sacrifice.

We again emphasize that these misperceived experiences, included giving self away and allowing energy feeding dependencies, with nothing left for self. It's hard for life to fulfill and serve you when you keep **<u>splitting off from self, without caring for and loving self, first</u>**. False love in the mind always justified it as the human years go/went by with daily obligations, as you lost more and more of self. But that means you never really choose yourself and meant it, because trauma blocked your DNA divine memory of who you really were. Then you quietly end up sabotaging self and fighting life, fighting the body, fighting the mind, and being terrorized by the different wounded cast of characters within: Mrs. Angry, Little Ms disillusioned, little Ms Fear inside her shame and guilt, Mr. Self -Destruct and then you accept other's projections of who they think you should be. All the while underneath these projections ***your natural innocent joy holds its breath.*** Then, thinking yourself unlovable and unworthy, while still holding back your core soul light, your children and your inner child, feel this contamination. Parental Wounds/patterns run deep and must be scrubbed at the DNA level. This causes death, disease and needless suffering as you all now know. Time to come home to yourself and your children will awaken also and find their true path. They are waiting for you to love yourself. And just when you thought you might get free, the one you projected all our love and wounds into, gets infirm. And you end up taking care of them, their life's mess, penniless, as they lay dying. Then your dignity is trapped again in the old illusion and fear like an energy feeding virus eating your wonderful empathy in the form of an elusive lover who has won again. And you must be born and die again to you. This is because you didn't allow your heart to again open fully **to your Core Presence.**

In the **Old Earth Heart**, just finding the inner or outer relationship of a loving goddess, a loving friend, a loving Mother God restored the core of love's existence until the next traumatic human fear revisited from the past programming in the DNA. In this transition the old sensitivities want to transform into new communication senses. What if this old different wounded **cast of characters within** were given new **fulfilling roles**? When she/he went all the way through the core light, Mr. Angry became a hot pink passion candy on the tongue. Little Ms. disillusioned had so much awakened energy she never slept and only regenerated in the heart chamber twice a day. Mr. Big Fear inside his shame and guilt shed his mosquito shell into a Monarch butterfly wing for his new website logo. Mr. Self- Destruct, set up a world program to change all prisons into re-education schools. And, Bully Boy turned in his boxing gloves for a dog training school that could smell disease pathogens before they created epidemics. Indeed, some of you are ready to go all the way thru the old pain body sensitivities in your final transfiguration to access your new master code communication light systems.

Indeed, then LOVE has been redemptive and restored. The old heart's pain body had wounding emotional qualities of love such as: fear, vulnerability, hurt, betrayal, loss, abuse, guilt, and shame. The new bio-heart's compassion wants Love to take the

pain body and grow it into new qualities of expression. *Those can be the gifts now from Mother Earth love as well as, each soul's divine god/goddess nurturing within new essence love qualities of: joy, new passions, renewed playfulness, fulfillment, transparency, core trust and inner communication.* Nothing is taken or given in love. It always was and will be the Eternal Presence of Life, but will keep growing new DNA helixes and new species qualities. Yes, new kinds of love will now evolve through the evolution of consciousness because of the heroes' journey and genetic **experiment of the** *Gaia Soul-Spirit of humanity. Merging and rebirthing the human back inside the __newly born spirit heart__ has afforded this opportunity. And that is also why you will no longer be able to able to expect, control how or if; or even be able to say that another isn't enough love for you. This is simply the old protective heart avoiding hurt. All you can do is be present and stay, stay, in that beautiful empty space in the heart and exchange what your passion is with another. And when that is shared, each soul in that very moment has created an instant manifest; and via communication, free energy transmits it to all existence. Each soul is then free to accept or receive it within their composite vibration; the true value of their essence self. You then, only speak or telepath from the new heart voice. Others can sense that you are no longer a victim of fear. You illuminate that you love yourself enough to finish healing all aspects of self through the Core Being of your heart. This enhances and gifts your vulnerable intuitive sensitivities to create new qualities of joy, instead of feeling held hostage by these human senses.* Others even sense that if they dare open the heart a little; somehow that it won't have to diminish their worth or self-value. Remember the lower vibrations of humanity's perceptions can only view each other through their own filters and patterns if they are un-awakened to their heart light.

Your consciousness created you from a natural state inside the essence of Love's joy; an energy expression substance and quality of existence. You took it through a journey of wound, depression of energy, lack, and constant challenge to return home as the hero of your own story. When you return to your new heart home, you hear a new song playing. Your inner master code communication from spirit says to you. *"I love the way we can live together in our moment in that empty space in our heart. I love the way we trans-sense, and are intimate with the images, potentials, visions, expressions of our next instant manifestation. Our manifest always wants to answer to our new passions, realizations, fulfilling real joys that arise within our new Eternal Sovereign Soul Presence Potential. The cosmos communicates through our new heart's excitement or passion sense about whatever we are willing to accept. Our heart vibrates information of light inside our new network light codes in our new DNA that excite our new qualities of love. All we have to do is sit together in the heart chamber in our new Divine Bio-Human vessel and the cosmos will transmit the image, new quality sense or re-imprint our DNA into whatever experience we want to become."*

Indeed, all your old colors, light, sound languages have expanded their mathematics, science, and technologies through this journey. Creations, solutions, and community are always found via simple transmission of hearts as each receives at their own vibration what is suitable in the moment of that soul! Eternal love qualities of joy, new passions, renewed playfulness, fulfillment, transparency offer new octaves of conscious vibration. There is always gravity in the heart's magnetic DNA cells to guide what matter imprints in the gravity of each situation. Nothing is taken or given in love. Again, there is only communication and equanimity of exchange. Love is again evolving itself through the evolution of unique embodied soul consciousness.

The Myth of Forgiveness Ascends 7 -2019

Bio-Light Masters, forgiveness will ascend back into **Divine memory**. And, as you reenter the Infinity Star Gates of quantum multi-light energy fields, you are creating in the new energies. How did these energies evolve for use in new applications? If you view your universe through the eyes of self- realization and freedom's love, rather than the old Earth energy of extreme polarity; then the truth of humanity Divine origins reveals itself. Rather than the brutal savages' humanity has been purported to be by, ancient misguided-antiquated creator-god beings, who wished to colonize Earth for its precious minerals, DNA genetic laboratory, or Homo Sapiens' labor force; the truth is that the human Spirit has provided a new bio-heart imprint or template for all creation. Indeed, Creator Earth School continues to balance and solve ancient distortions in order that all cosmic beings better understand the free will DNA codes of evolution in service to all life. This included the choice of becoming a humanoid soul essence being within your local universe in order to become your own conscious soul sovereign creator.

So, when you came forth into this universe, Creation asked that you help build this local universe and then experience every possible potential within it. You did this through the archetypal roles with your spirit, soul, and human families by exchanging all possible roles, lifetimes, and essence quality expressions within your DNA blueprints. This included being everything from: a particle DNA-cell code for a universe; to a galactic warrior, an energy wave band of light, or a human mom, dad, or child. These pattern codes could be energy or dense forms, sense expressions, existences or lifetimes; or any life form potential you could imagine. In this Universal school of potentials, you have mastered as creators: marriage and family, genetic free will, space-time atoms and quantum energy exchanges in new cosmic matter. You have even lived all the various creation Myths. This included any distortions of trapping, forcing, controlling, or abusing life's evolution that arose as free will choices began to become distorted by **other creators.** For example, there were those interfering and forcing evolution from outside your universe with different agendas or DNA codes to promulgate a slave-trade and mind control right to rule bleed through, as mental simulations. You all sensed this as a kind of distracted confusion between creations' Ancient Founders and the Original Essence codes' architectural blueprints mixed with mental competition from those outside your universe and outside your DNA blueprints internal communication systems. This caused you to question the worth of your existence and whether or not you should have left the comfort and safety of Father/Mother's Creator's love to further evolve or descend into existence in this universe, which you were designing through direct experience. This miscommunication caused you to go outside your heart Essence and listen to an external mind voice. This was called the 'Fall' or separation from Cosmic Heart. This is also where most of you bonded more with the Mother or Father Creator parent just as you did in this life. This original miscommunication doubt/judgment with your spirit heart about your right to exist caused one of the first separations from your own direct experience and embedded in all your creations henceforth. This was a misperception because **free will sanctioned all experience**. There could be no violation in any reality. Why? Because, what you created for yourself lived only <u>within your own consciousness</u>, thereby within your own creation, even though you shared a group consciousness. Hence it was impossible to harm self or another. It was only by: perception, judgment, or projection, that each creator could misunderstand how infinite waves inside particle light could descend the atom into extreme polarized cause and effect fragmentations within spirit's families. The atom could easily balance and adapt to evolutions growth at 2% polarity, but at 90% judgments' perceptions, fragmented the DNA stranding and information networks in the cosmic morphogenetic energy fields. Creators would henceforth have to learn how to stabilize their core essence from further fragmenting its experience in order to keep the fabric of particle matter together. You have all learned well; that if you fragment an experience and don't stay in the heart's original coded imprint to change it when fulfilled by it; then the human mind will try to: doubt, blame, shame, judge, fight, or fear its way out of it. Such a reaction: cuts off, traps, hides, or tries to control the life force energy. <u>Energy</u> always seeks balance, is always free, and life does not trap, force, or control energy in any way. Conscious evolution is free energy expression in experiential soul qualities of essence love. Before you doubted your existence, every polarized experience could be immediately balanced by changing the essence imprint within your consciousness. You could dissolve an imprint back into free energy and create a new imprint in any color, quality sense, or sound of your DNA rainbow field at will. Hence, **<u>no experience ever required forgiveness</u>**. Why would you need forgiveness for experiences you CHOSE to grow your soul to express the Essence coded in your natural DNA; even if studying possible separation from self was part of that experience? No sacrifice was ever required. The DNA-code design was for the direct experience. Any imbalanced, reactive, or restrictive DNA experiences were supposed to offer change or re-imprinted responses within the unique soul and it families. This would have allowed the essence extracted to sense new qualities of love's nuanced potentials to share with the group soul agreements. The choice began to present as feeling trapped in miss qualified energy. This did not allow **core essence light to serve all life** by continually treating every experience and each soul resonance as valued equally by Creation. This is also true of your Divine RNA Male and DNA Female that had different roles, but was always the same being within you.

So, in your ascendant awareness, you now can again remember and sense, that if you change the slightest choice in your moment all other potentials change; always seeking the most fulfilling choice. Indeed, it your evolving nature to: accept, allow, change, re-imprint, regenerate, fuse, re-essence, or re-molecule any experience you want. This is done through your new Heart imprint. You have remembered that you are all now living inside the Presence of your own new creations. Creation always marries the infinite and the quantum to birth new life. Creation doesn't question its existence or make mistakes and neither did you. Any needs for forgiveness are replaced by Divine memory. Forgiveness is built into free will and was never needed, except through mental control systems, because free energy or the life force energy is the natural communication with all life's fulfillment, and you are all life. If you make a request or command onto life, the answer is always yes! If you are descending energy into matter, then the vibration is slower. If you're ascending matter into energy, then the vibration is faster. If you ask that any fragmented or split off aspects of soul, lives, atoms, or spaces of your creation be returned to you, then they must; because, you are their creator. Or, they are sent to the source that created them. Creation naturally fulfils all potentials of its very essence expression in evolving eternal realizations. As a new bio-genome life form or

Eternal Being, passing though the Infinity gates into the New Earth realms, you are *growing endless qualities of love into infinite senses*. If divine memory inside freedom of self- realized love is truly transformative through the direct experience of each unique soul, then Little Boy and Little Girl Fear are no longer trapped or imprisoned in the human mind and can grow their heart into new qualities of love. Then **fear itself;** or the **anti-life shadow** within you that dies over and over to self, is also loved for its service to life and all experience is equalized inside all love's creations. Even fear can reach another potential. Perhaps it will become a new passion. At Life's core, even the most heart wrenching experience is joy, because existence takes joy in just being alive and the divine right to express anything. This means that forgiveness no longer justifies love in the mind, but remembers and opens a direct path through the heart and can re-imprint new qualities of love. This is the way of the heart. This means that your spirit's infinite love is no longer disguised by old human passions but alive, passionate, trans-sensual, and can be authentically lived in every situation without exception. It also means that even fear and forgiveness maximize their potentials inside loves unique passion of each heart, as you inform creation of its own growth. The Presence of Life can embody in any form and now the soul of humanity has access to **change its future potentials** in ways never dreamed possible, due to your pioneering love.

So, if everything is alive and the soul heart can truly re-imprint essence; then water can be changed into wine before the public eye as a real experience. Thus, every organism has a heart and is its own anti-pathogenic life changer. This is because it is a living truth that your spirit heart can change matter into energy though the essence of the soul's consciousness. Hence, two or more of you can be your own moment in your own creation sphere and still share a heart experience. This holds true even though you can be in different frequencies of vibration with another. Each receives their own choice within the multi-energy band widths of quantum cosmic rainbow spheres of free energy communication. Thus, each creates or shares in a creation without any interference in the other's soul path. Such a heart equation would equate incalculable potentials in new quality tones of joy. Such has been your path of the memory Divine; that you are all unique Hearts in a school to create a new prototype of Divine Human. When a critical mass of humanity remembers or realizes they are Earth's love stewards and Earth is theirs; their threshold of consciousness has the potential of healing your planet's vessel and perfecting their own vessels within weeks. Each density they ascend to brings that consensus reality closer and closer. Again, there is no such thing as death, only dead anti-life, or trapped energy held in distortion masking essence wave form consciousness of particle light in Creation's evolving potentials.

So, where has **Creator School led you**? You now live in new conscious communication bio-light network spheres within new qualities of infinite potentials to explore. You've self-realized that the divine memory of your soul heart's, new tonal-Heart signature-DNA imprint, has fully activated itself in your new cycle of evolution. This organic imprint has been upgraded by the blend of your human and divine senses merging into an upgraded super-sense intuition of heart's fulfillment, and its bio-luminescent transmission to All Life. And free energy replaces forgiveness. This is because free energy Essence lives in the Presence of all Life. Life's Presence is in an energy relationship which is alive, passionate, expressive, sensual and full of intimate communication. Consciousness is sense-essence energy expression including all the qualities grown by the soul's heart love through both human emotions and divine senses merging. And energy always seeks balance. Your discourses with your spirit answer that your new bio-conscious DNA light network is the new living code for your new intimate relationship with all matter/s of life. Your bio-organism's DNA talks to the cosmos and the cosmos talks back in the form of instant communication and direct manifestation. Again, and again we say, that your new multi or trans-sense communication with your inner Presence and all life is the master's code for the new ascending heart DNA transmissions.

Over and over, the cosmos hears your *__elegant voices__* of inner master code communication from your Spirit Heart *"I love the way we can live together in our moment in that empty space in our heart. I* *love the way we trans-sense, and are intimate with the images, potentials, visions, expressions of our next instant manifestation. Our manifest always wants to answer to our new passions, realizations, fulfilling real joys that arise within our new Eternal Sovereign Soul Presence Potential. The cosmos communicates through your new heart's excitement or passion sense about whatever you are willing to accept, change, image, or re-imprint. Heart vibrates information of consciousness inside our new network light codes in our new DNA that excite our new qualities of love. All you have to do is sit together in the heart chamber in our new Divine Bio-Human vessel and the cosmos will transmit the image, new quality sense or re-imprint our DNA into whatever experience you want to become."* Indeed, conscious evolution is free energy expression, in ever growing soul qualities of love, allowing the heart to create anything out of pure essence. Direct experiences in your multi-light heart's path needs no forgiveness, because it was always the way.

Bio-Ascendant Contact 2019-2021 with Star Families 8-2019

Quantum Creational Bio-masters, you now begin to set the pace through a template of direct experience for inner and outer contact. You do this by creating through your super consciousness gifts or trans-soma Essences and this is **why you are here at this time**. As bio-human masters, this bio-luminescence transmission is a great gift to stabilize humanity. This is important now, since the New Earth Star-Sun soul gates are open for migration choices into the New Earth Universes.

So, the <u>wonderful news</u> that most of you must be reminded of, is the self- realization that you had one or more aspects that had already graduated from the **parallel ascended universe**. Here there was no alien ET intervention, no fallen angels, or bio-light distortions. You mastered the universe according to your natural essence codes. Your existences explored attribute expressions of beauty, joy, and all multi- essence senses. You grew and manifested new life via their direct experience. These were aspects, facets, and existences, of lifetimes of both inter-dimensional and dimensional expressions.

Again, each attribute or facet was like a new flower, a new organism, an existence, a helix, anew ecology which informed the cosmos what is was becoming through its own creations. In the multi-verse, these were inside your natural state of being and you came to experience every possible potential expression of these sense attributes of light, color, and sound qualities and transmit them throughout the multi-verse and the cosmos. **Even then**, they were as new essences folding or states of streaming new consciousness. And, your consciousness then allowed love's emergence to become itself again and again and be the substance of its own manifestation.

However, when you **came into this universe,** (from a potential future), to create a bio-physical free will universe, you had to become the distortions of: the misuse of energy, the extreme polarities, the DNA violations by alien worlds and agendas in order to transmute them and assist. This meant you allowed your advanced **multi-consciousness to lay dormant,** until you were sure all your creations in this local universe, would have the template for their enlightenment and bio-ascension. As a human who built a body of experience for the soul, you appeared stuck in an addictive reincarnation cycle. This often made you feel like you had to hold back your core light and natural essence senses in order to master co-creation. Now your self- realization and self -awareness have channeled back, to tell you it's time to return home to the multi-verse.

But, **since you left, multi-verse has grown new essence DNA attributes and new qualities** from the wisdom of all the aspects that you lived in the Old Earth Universe; where you trained to become your own unique soul's creator and master the atom's space time matter. As you return to your own bio-creation and move into the New Earth realms, *you may interact* with conclaves of beings coming in to support the ascension. Some of these are still in the higher realms or spheres of the Old Earth Universe and others may be coming in from the new Super Universe, just as you came from the Multi-verse. And many of them are here with you now if you allow your trans-soma heart to sense them. You may interact or channel them in various ways. They can come into your heart chamber channel via: video streams, dream realities, inspirations, invention ideas, direct manifestations; or through your beautiful new quantum senses and code fractals. These star families can manifest into form as required. And, most are already all over your world assisting in the transition helping you to channel what humanity most needs to regenerate their planet, and for them. They are you and you are them. They are either members of your Old Earth star families or your new quantum star families. They are already carrying the new species DNA you transmitted to them from your sojourns in a past-future space-time universe. You have grown your love's essences into a new cosmic cycle, with a new Divine Human free will prototype that will enable creation to secure and create more bio-physical universes. This is, such that the non-physical **realms can choose to unique a DNA bio-soul** as their own creator versus being limited to the non-physical oneness as the only expression.

So, ascendants, what are the **new attribute expressions** of: beauty, collaboration, compassion, creative passion, etc., and how do they grow and manifest themselves through your experiences? Again, each attribute is like a new flower, a new organism, an existence, a helix, anew ecology which informs the cosmos what is becoming through its own creations. Consciousness then allows your Eternal Presence to embody or become itself through **the essence substance of its own manifestation**. Are you ready to allow sharing your essence attributes and their realizations through the heart's guidance system? Hence, you can create any version of New Earth existence you wish to experience, including both inner and outer contact with your star families. And, as a Creational Master, you can enjoy any of the trillions of scenarios and choices ascended living now avails! We reiterate that all potentials, realms, or realities are created inside the **heart chamber's** elegant love codes. The adventure of being an integrated Sovereign Divine Human Master in the continuum of the new quantum model ascension vessel, offer all kinds of wild choices. Allow your creative imagination to again be your true natural essence reality.

Unleashing Potentials as Judgment Ascends into Divine Justice 9-2019

Bio-quantum Masters, new potentials are unleashing through you, as Judgment ascends into Divine Justice. Let's follow this potential and track its ongoing effects on you and your worlds; so, you can Presence **how you're creating these new potentials**. There is a vast difference being revealed in your realization between unleashing potentials in awareness or judgment-perceptions. Indeed, the new bio-essence DNA-heart compassion has grown judgment into a new essence quality of Justice. Such is free energy communication. How? Many argue, or commune within Core Self, as to how you can keep a pulse or assist in the world **without being pulled into its chaos**. This chaos by order is bringing about massive change to transform all old dinosaur systems that humanity has outgrown in their evolution. Cosmic Bio-physics does not allow evolution to go against life. Life does not anti-life itself or de-evolve. Hence, with your new super intuitive consciousness, you can bring your attention to any old trapped energy systems such as: politics, religion, technology or human suffering and use your Master awareness that offers higher frequency potentials for benevolent outcomes. This realization keeps you from being an energy filtering empathic virus for all the world's conflicts. Your consciousness, simply through allowing observation or awareness, will transmit a higher potential to all of humanity; such that, a higher frequency choice can be sensed throughout the planet's resonance. Awareness is one of your finest essence/sense instruments. It always brings an opening for human spirits to share soul heart's openings for innovations, solutions, communities, and unity. This awareness amplifies and neutralizes any imbalanced or miss-qualified energy such as: war genocide, derision, or paralyzing argument. Communication in the transparency of the heart avails its own ethic and reveals unseen potentials felt throughout your world and informs the cosmos. Truth is subjective experience and when all views and choices are shared, new windows and opportunities are no longer hidden by agenda.

If you chart your soul's progress you self-realize that its unique growth has always been in perfect equilibrium with your soul and the cosmos. You have **walked the embodied journey** by mastering human lifetimes and soul-spirit existences through **energy rainbow bands** in the light spectrum or pyramid of atomic light. These frequencies pass through your Essence stem cells-DNA information and bio-spheres of consciousness in constant flux. As we climb this pyramid of light, it is important to mention that you can't deprogram or change an experience from the frequency in which it was created. It must be from a <u>higher vibration</u> or the next octave of light. So, first, you deprogrammed mind-emotions recycling dense subconscious programs. These were hertz, infrared, and visible light frequencies. And remember, information is not stored in the brain computer, only in the DNA; but the brain rewires and processes it out via electro-chemical neural responses. Then you moved to re-program collective unconscious human feelings within personal thoughts, feelings, attitudes, and beliefs. This was mostly visible light, and spiking ultra-violet frequencies. Next, you remembered your angelic soul senses and those pattern aspects/lifetimes of self and soul group. This was usually ultraviolet and X-ray frequencies. Then, your co-creator spirit re-embodied in your core light essence to reintegrate all of your human-soul-spirit stories, movies, and images for their wisdom. And this included your oversoul group, which spawned you forth for this creation. These are/were gamma and cosmic ray frequency vibrations in your Pyramid of atomic light. This triggered the pyramid of light to **spin like a Bio-ship/torsion field,** (or a rainbow spiral), as you continued in your enlightenment to move beyond and lift the veils of mind, geometry, space, time in order to re-connect to core creation again. This released the old ancestral DNA and its programming; that had allowed for all limited potentials to be experienced, through all 144 densities with the 144 dimensions within all archetypes in your local universe. This would prepare for light body to receive enlightened frequencies and the new Essence-DNA energy imprint through receiving the quantum frequency rainbow spheres for ascension. *This is why ascended masters appear in robes, as their* *span represents their own unique energy field or bio-spheres, in which they can fly or change/imprint their essence into any life form.* And yes, if you are still in the old body DNA networks, the brain, and mind-emotions; then you may need to feed your human sac of chemicals, neuron-transmitters, hormones, and bones with: adrenals, medicines, pills, and doctors. This electro-chemical process recycles until the soul crystal cell, the diamond spirit cell, and plasma particle cells awakens/re-splices the DNA; to prepare dormant DNA master human activation.

Master Essence core-DNA, (stem cell), then activates the dormant new Essence species-DNA quantum frequencies gradually. This grows your newly born bio-essence Homo sapiens DNA-heart imprint. This is a composite blend of all old energy limited essence quality potentials such as angry mind, hurt heart, dying body. Mastery of these will be blended into quantum frequencies and unlimited DNA- essence qualities such as instant manifestation and super-sense knowing. This new bio-imprint contains a diverse composite of the entire DNA that has ever been created in the cosmos and serves as a new upgraded Homo sapiens soul-spirit imprint essence. Its quantum qualities in its light, color, and sound codes are: conscious/self-knowing, translocation, trans-soma creation, and magnetic vacuum gravity acoustics. As we have said again and again, it is also the imprint for each unique soul sovereign new bio-Heart-DNA Essence vessel. It can re-imprint, change or create any life form. This is the next prototype for Divine human DNA which has its own individual morphogenetic biosphere, or energy field, which can grow unlimited types of DNA for future generations of star beings in the cosmos.

Hence, with your new super intuitive consciousness; your natural **consciousness of justice** simply releases old trapped energy systems such as: politics, religion, technology or human suffering. Just by being fully embodied; your Master awareness that pioneers and transmits quantum frequencies for benevolent potential outcomes, offers change to those who accept it. This realization keeps you from being an energy filtering empathic virus for all the **world's conflicts**. Your new DNA quantum consciousness frequencies will transmit higher potentials to all of humanity; such that, a higher frequency choice can be sensed throughout the planet. For those trapped in the mind, havoc in the body jiggles the soul to try to awaken. *Hence, your new bio-matter DNA-code communication free energy network is your finest essence/sense instrument. Your code organically knows, that when you do no harm to self, it is impossible to harm any form of life. And when you heal yourself, you automatically heal the planet.* This open communication always brings an opening for human spirits to share soul awareness openings for innovations, solutions, communities, and unity. It comes from a human master with **Sovereign Soul Essence** which answers to free energy communication of new essence passions, realizations fulfilling real joys. Their life's passion fulfillment answers only to moment to moment potentials that can change or re-imprint any life form. This human master awareness amplifies benevolent choice potentials for all, while neutralizing any imbalanced or miss-qualified energy of war, derision, and argument. Master code communication in the transparency of the Essence DNA- heart **avails its own ethic** and reveals unseen potentials felt throughout your world and informs the cosmos of its changing growth.

Truth is subjective experience and when all views and choices are shared equally, new windows and opportunities are no longer hidden by agendas locked in prisons of mind, power, and control that intimidates or harms others. These old energies also **bind the controllers** in their minds and technology unable to access essence consciousness. They end up fighting others of like vibration and themselves and cannot sense the higher frequency potentials your transmissions are offering them to go beyond the mind and sense Essence. Enlightenment requires you lose/reprogram your mind-emotions to go beyond them. Enlightenment is impossible through the mind. It comes through the Essence DNA-heart. Hence, no one leaves you in depression or loss, and you don't leave anyone in anger or hurt. Each of you **just owns more of yourselves** from the reflection of shared experience. Sovereign is love's essence and that is why you never lose anyone or anything. You just get a deeper sense essence expression of your Presence in reflection and absorption of more of what you already are. But that sweet, tasty sense of intimate, fulfilled, intimate communication with life's actors shared between souls; imprints all life with one impeccable, indelible glance as a new potential. And, in your enlightenments you've had to: channel, write, teach, and re-grow your souls out of the blind crazy mind to sing, dance, and play your own new soul essence song again!

How again did **judgment potential justice?** Human's new bio-DNA species essence compassion has grown the old Earth pain body of fear, loss, guilt, and limitation into new essence-sense qualities of joy, passion, fulfillment, transparency and inner communication. All trapped or extreme energies can be **released into new potentials**, because the new bio-essence DNA-codes now operates within a free energy light communication system of your own Master Presence within all life. Militaries and governments can open free energy technologies for peace. Cultured animal, plant, and crystal essence cells can be **incubated or re-imprinted** for new food sources without killing life force. The master code stem cell re-splices/regenerates the essence DNA to heal any disease or re-grow an organ. Organic Cyber sensor-chips can answer to the bio-organics consciousness of the human species, to avert disasters, cyber entrapment, or genocide. So, what is this free energy essence expression that communicates in service to all life? Free energy lives in the Essence Presence of all Life. Life's Presence is in an energy relationship which is alive, passionate, expressive, sensual and full of intimate essence communication within the natural essence of each unique soul.

In applications of <u>personal mastery</u> with no mind control programming, your awareness/self- realizations allow new experiences, as your trans-sense your essence into new qualities of bio-cell love. It's simply living in your new unique bio-awakened Spirit Presence. When <u>uncomfortable</u> in any way, free energy communication always reminds you to re- stabilize your light core Presence and shift completely into your Heart Presence/chamber, which is your consciousness, intuit heart awareness, or the Present moment. Then, as you take care of self needs, passions, or creations first in every moment, then you're able to sense another's program trying to project on to your light or be drawn to your reflection. The human master is **unaffected** by outside projection because their DNA-essence energy field will respond according to the highest potential of vibratory exchange or transmission. Your composite Spirit Presence always neutralizes or harmonizes the energies by creating your passion potential for you first. Then, communication transmits to another according to their soul's agenda with harm to none. Direct spirit communication is always transmitted if you are first sitting inside your own soul essence-DNA imprint. This is because the bio=physics of an awakened energy field no longer accepts another's self-judgments, self-rejections or self -abandonments. You also know that as family, friends, and the past is released over and over, the human mind-emotions can feel like a great loss if it has no friends or clan or soul group to replace them with. This has been because there has been such a strong human need to fit in and conform as a false mind sense of grounding to life, which replaced the natural **internal essence communication**. However, growing wisdom in self-realizations in your enlightened years has told you, that family and friends were just human-soul mirror reflections of our own inner aspects like anyone else. Without all the drama; Mom is just mom, brother is just being brother, and life is just theater where they must finish their soul stories. You just enjoy them as they are, because there is no longer an electrical chemical cell-charge on what they say; as your human heart's bag of chemical-emotions can no longer be hurt. And you have re-integrated any lost fragments of SELF. It's all just Life and its props and each being belongs to their own soul-spirit. So, all this Self-realization awareness and movie memory of your story, is your spirit guiding you in back into free essence energy communication, bringing that which serves your potentials first. Then it automatically transmits abase alpha wave consciousness to the soul of your world, **sealing out any mind control**.

This clears the way for **direct ascendant experiences** that you have as master code creators such that; your manifest always answers to our new passions, realizations, fulfilling real joys that arise within our new Eternal Sovereign Soul Presence Potential.

Again, cosmos communicates through new passion trans-senses about whatever essences you are willing to accept, change, image, or re-imprint. And, *you change nothing except through your essence*. Thus, divine justice is free energy essence expression in equilibrium and equality of worth. When you make a request unto life, the answer is always yes! Life does not reject or abandon itself. Consciousness is sense-essence energy expression including all the qualities grown by the soul's DNA-Essence qualities. Free Energy lives in the Presence of all Life's existence. Everything is energy, and energy is everywhere. Life is free energy or passion in motion. Your new vessel imprint is a bio-DNA-essence conscious animation of energy inside qualities of your own spaces, time, and gravity within your own conscious **creation sphere.** And every particle of life has its **own gravity,** otherwise you could not unique and anchor the core of your new soul Presence within your own creations. Hence, this free energy sovereignty of a human master is divine justice because it allows each soul the opportunity to ascend judgment/ or separation from Self, into a new potential. And, part of **ascension fulfillment**, is for each soul to experience the effect of all potential outcomes that this multi- light cycle consciousness has, on themselves and next generations in your diverse genetic experiment; where civilizations from all over the cosmos are represented.

Are you beginning to notice the dramatic changes the **effect of your master consciousness** is having on your world? It will become more and more apparent how humanity is less accepting of any form of human slavery that harms the soul-spirit. This stewardship of the heart-soul of each other and your planet can ameliorate anything on your world, until any injustice is no longer acceptable. If humanity reaches a consensus reality on any issue whatsoever, then it released from your planet and reported in the news the next day. You are still here to witness, transmit, and enjoy being the pioneers of these new realities as they fulfill your creations also! Humanity calls them miracles, but they were just untapped potentials.

So, in your ascendant awareness, you now can again remember and sense, that if you change the slightest choice in your moment all other potentials change; always seeking the most fulfilling choice. We repeat again and again, that it is your *evolving creative nature to: accept, allow, change, re-imprint, regenerate, fuse, re-essence, or re-molecule any experience you want.* This is done through your new DNA soul Essence imprint. Now you know why we have called your **heart chamber** *a: DNA- Essence Potential Source code/r, a bio-organic essence magnet, an atomic fusion accelerator, a stem cell replicator and regenerator, a quark centrifuge, a morphogenetic torsion field, and a midwife of worlds. Indeed, it must be Essence/d and lived to be imprinted.* You have realized that you're now living inside your unique soul's Multi-Presence of your own new creations. Creation always marries the infinite and the quantum to birth new life. Creation doesn't question its existence or make mistakes and neither did you. *The Presence of* **Life's** *existence is its own justice and equalizer.* **Only judgment can abandon itself,** until Essence DNA soul awareness loves it back into new free will choices, within divine justice and the right to exist. Divine justice has always lived inside your DNA essence master code for equal energy expression and communication with souls and all life. You have reawakened divine justice in a new day to day potential, **giving humanity hope** that new ways of living in multi-consciousness are now available.

Finally, have you not also been designing your own creation to feel and sense all its potentials? This has allowed you to invent multi-variant ways a Creation is built and architected. How many solar systems, stars, planets, or suns does your new creation have? What is its geometry and light, color, and sound frequency parameter? What species DNA codes/patterns, or civilizations and life forms occupy it? What levels of consciousness is your creation in harmony with? You have simulated destroying life and trying to change it in many forms in order to express different levels of consciousness in cause-effect atomic density and quantum quark energy matter. Indeed, you have come to Creator School to learn how to create whatever you desire from the direct essence experience of the growth of your soul's wisdom, but now as your own Sovereign Human Master. This is because you have wisdom or direct experiential knowing; that creation allows you to create an entire universe, experience it as if it is real and dissolve it back into pure conscious essence energy; as a: video story, a simulation, an illusion, as real, imagination; or as if it never happened at all. But you take with you unlimited **wisdom combinations of unleashing potentials;** in **new essence imprints** that will become new qualities of sensing energy, communication and imprinting life's evolutions throughout the cosmos. In fact, could you fold or imprint the entire cosmos into **one DNA-cell** of consciousness? One cell of imagination?

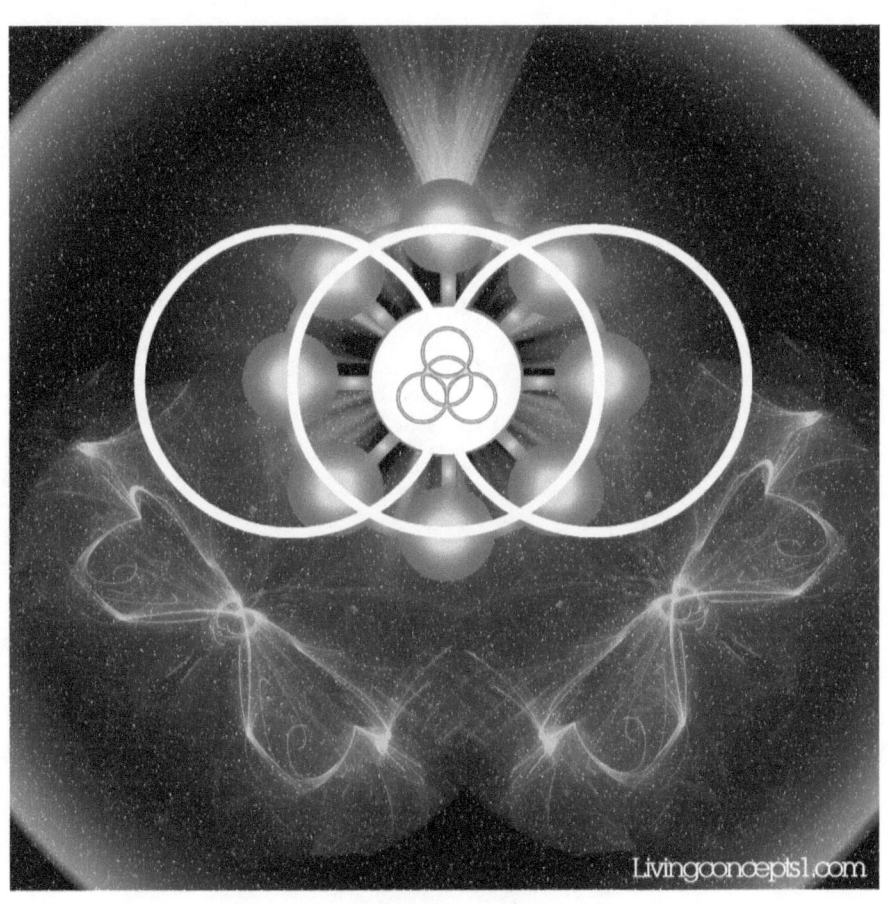

Free Energy Creation Reminders 10- 2019
Q: Can you review some basic guidelines for the Master Code Creative free energy process?

Bio-Light Masters, today we review the main creative growth stages of your evolving soul-spirit Essence as a Sovereign Free Energy Human Master. This will help you be poised for the rapid acceleration in your creative energies that will change you, your planet, and the entire cosmos in the next decade. As, you ascend daily into the free energy essence heart-DNA, you now understand that you're not just your body. The *vessel is a bio-imprint of the heart's DNA Essence*. Your creative energy, your life force energy, and your sexual energy are all the same free Essence energy that makes up your consciousness. The cosmic fire elements of the: IAM soul, the IAMTHATIAM Spirit, and the Eternal Presence, embodied as Divine-Human Free Energy Master; are your Essence consciousness and the Essence expression of all life.

The sacred body shroud, imprint, or animation is just a tiny light factor of spirit essence. You simply can't fit all your consciousness inside the vessel and that is why the Master Essence can change the Heart's DNA-Bio- imprint into any life form chosen. This is because the DNA contains all the information of essence imprinting, not the brain or the body. They operate as a computer inside a bio-organic Essence cell pod of electro-chemicals. The brain-body's past-future memory video programming is gradually decoded and detoxed, throughout the cell embody. This happens while merging aspects, lifetimes, or existences of the soul, and also in spirit oversoul merges or rebirths. There is constant continual bio-integration as each octave of light is achieved. Essence heart-DNA changes the body, and your sac of human electro-chemicals responds as you raise you overall soul's light quotient. Again, the mind-emotion brain is a computer to process programmed memory systems for the human bio-organic chemicals.

The **progression** to **grow new spirit essence** has been from: subconscious mind-emotions to>human feelings to>angelic senses> to Human-Divine spirit Essence senses; growing into new multi-quantum qualities of rainbow spheres; or new qualities of essence love. In a natural state, the composite embodied soul-spirit **Essence** changes experiences moment to moment, once they are fulfilled. They were never meant to be: programmed, recycled, memorized, recycled, or re-incarnated. Only Heart's self-awareness or self- realization can truly change your reality. Divine memory tells you Master Creators, as you watch the illusions of your old Earth fade; why nothing exists except what is inside your own consciousness. The **Presence of Self Love**, awareness, self- realization, or conscious potential; is created from an awakened DNA-heart, that can sense-intuit and communicate any experience inside itself, and smack its lips with the essence of its own fulfillment. In this consciousness life, is an intimately unique trans-sense experience, because you remember that you created and are, naturally in a continual relationship with all life forms. Natural Love's essence is a refined silken essence substance, within beautiful qualities of attribute expressions like the cinnamon laced petals on a rose.

Indeed, this intimate communication with SELF in these intimate experiences with life grows the substance matter of consciousness via **direct Essence sense experience.** And, the substance matter or reality takes the form of life and that includes any blueprint for any life form. There is nothing in this or any world you existed in, *that you really change, except your own sense essence qualities of fulfillment within each moment of experience.* The color blue can be: the blue-red sky, a wax crayon, hopeful inspiration, sad violin notes, the moon's liquid light water; or take on any quality of multi-light, color or sound. Dense or quantum qualities of Essence design our expressions.

Inner heart channel is able to communicate on all levels of existence at once to receive any expression that fulfills each soul in each Present moment. This is instant manifestation right out of essence consciousness, and access to all other realms of existence in quantum applications. Soul is enlightened and the spirit's ascent as a Human-Angel-Master Creator opens a new DNA-Essence on a new adventure life cycle. Restoration of Divine Memory communicates to you that everything is this life has come to be about your core light expressions as a unique soul Sovereign Human Master Creator. Each Master has their own bio-physics agenda for their natural progression. Bio-Light Masters trans-sense communication inside their own bio-light network morphogenetic spheres; or rainbow torsion fields.

Overall, Divine memory of your soul's composite tonal <u>DNA imprint</u> has fully activated itself in your new cycle of evolution, wherein humanity must be embodied in the crystalline diamond cell to overcome death and gain access to the New Earth star gates where enlightenment paths are offered. For others the, (quark-liquid-particle-plasma master cell), can be lived. This organic soul blueprint has been upgraded by the blend of your human and divine senses, merging into an upgraded imprint, inside **super-sense intuition of heart's fulfillment**; and its bio-luminescent transmission to All Life. This is activated, because your new conscious DNA communication **bio-light systems** are available and functional, for your full access. Hence, your new bio-conscious DNA light network is the <u>new living code</u> for the intimate relationship between consciousness, energy, and creating new (dark) matter. Your bio-organism's DNA talks to the cosmos and the cosmos talks back in the form of instant communication and direct manifestation.

Indeed, your new multi or **<u>trans-sense communication</u>** with your inner Presence and all life, acts as the **<u>master's code</u>** for ALL life and the new ascending heart DNA transmissions. These rainbow bio-spheres within spheres access new DNA codes, and their soul communication expressions, can't be influenced by external people, places, circumstances, or events. This is an internal communion with all existence in **<u>every now</u>**. And, each moment can be new. This new DNA heart-gate or bio conscious I-phone that you have become transmits the new standard of consciousness for all life cycles. This is a bio- light network communication inside your heart channel in/on all realms of existence at once, without interference from any external reality whatsoever! Hence each moment inside your PRESENCE is fulfillment as its own occurrence or manifestation. This makes the organic, Soul infused homo-sapiens' Essence, the most courageous and loving template to evolve in the cosmos thus far. Indeed, it's your time to remember the natural essence of bio-organic creation. Now that you are conscious you know that *Life always serves life no matter the experience.* Any perception is simply still a part of separation or distortion of experience; when it is understood that everything is pure consciousness.

Consciousness expresses via the essence of free energy. The <u>Life force</u> **Free energy** lives in the **Presence of all Life Imagination. Free energy is an intimate relationship with your own inner Presence and the Presence of All life.** <u>Creative free energy</u> is Life's Presence is in an energy relationship which is alive, passionate, expressive, sensual and full of intimate communication. Consciousness is sense-essence energy expression including all the qualities grown by the soul's DNA-heart love. Free Energy lives in the Presence of all Life's existence. Everything is energy, and energy is everywhere. *Life is free energy or passion in motion of imagination.* The new spirit heart is the **soul's DNA energy imprint.** All Life animates its Presence in the Essence expression of free energy. Even the sexual energy grows into energy relationships which are: awakened, alive, passionate, expressive, sensual and full of intimate soul-heart communication that grows the soul's spirit. Yes, life is free energy communication and energy always seeks balance. You vibrate matter up or down or at multiples of the speed of light, color, and sound. The new DNA Heart Presence Essence can **change anything** via the excitation/passion of particle light waves. The sum total of your Essence-DNA light quotient determines access to your morphogenetic field. If the bio-body/vessel is not treated as sacred and lives in the programming of negative thoughts, feelings, attitudes, or beliefs of being: sick, contaminated, cell toxic, or locked in an opioid drug brain; **this limits access to higher frequencies** of your own morphogenetic energy field, or creating and merging with new rainbow quantum energy fields.

Therefore, conscious evolution in free energy qualities of love produces new super-intuitive senses of life and **constant new potentials.** The cosmos communicates through your new heart's excitement or passion sense about whatever you are willing to accept, change, image, or re-imprint. *Hence, your **new bio-organism DNA genome is the anti-dote and immunity** for all: viral, cyber, or DNA pathogens; and is now able to absorb cosmic radiation and adjust any time slips or space warps with your own consciousness.* Indeed, you have restored any ancestral DNA bio-species distortions. Heart vibrates information of consciousness inside your new network light codes in your new DNA that excite your new qualities of love. Eternal love qualities of: joy, new passions, renewed playfulness, fulfillment, transparency offer new octaves of conscious vibration on moment to moment access. **Creation opens** to a level of intimate trans-sensual experience never experienced before. Imagination becomes its own anticipation!

There is always gravity in the heart's core essence magnet to guide what matter imprints in the gravity of each situation. Included are natural potentials such as being **your own free energy source.** You can change matter, remain in cell fusion and regeneration, and use tele-transmissions. You can you imprint and vibrate potentials in and out of different matter states of reality so that what matters to you is its own occurrence? You can travel on sound without need for a jet or beam from a spaceship. You can talk directly to souls or spirit beings without need for an I-phone as simulated consciousness? Your vessel is its own conscious technology and can imprint as any life form. So, in your ascendant awareness, you now can again remember and knowingly sense, that if you change the slightest choice in your moment all other **potentials change**; always initiating the most fulfilling choice. *Indeed, it your evolving nature to: accept, allow, change, re-imprint, regenerate, fuse, re-essence, or re-molecule any experience you want **because you have become it!** This is done through your new DNA Heart Essence imprint. Now you know why we have called your heart chamber a:* **DNA- Essence Potential** *Source code/r, a bio-organic essence magnet, an atomic fusion reactor, a stem cell replicator and regenerator, a quark centrifuge, a morphogenetic torsion field, and a midwife of worlds.*

So, your master code DNA-heart replicates or imprints any life form **via a heart pulse.** This happened because you transfigured the old limited human senses, or pain body, into new qualities of Divine Essence. **In the creative process,** your bio-organism simply follows its own codes and changes matter accordingly and this matter manifests in ongoing potentials as your reality. As your evolution changes, that also informs and changes the cosmos. You can dissolve any matter or substantiation of reality as no longer true, if it has been fulfilled, and energy no longer maintains or follows that focus. Or, you can remove your Love's perception and its complete focus from programmed, patterned, or fulfilled realities that your soul has outgrown. Then, these perceptions or subtle judgments no longer exist and disappear, or dissolve inside your conscious awareness back into pure essence or free energy. You return to the still Eternal Potential Presence of the heart chamber, where new communication in a new moment awaits! Here, you will always hear the song of spirit heart's master love code saying, *"I love the way our heart trans-sense, and are intimate with the images, potentials, visions, expressions of our next instant manifestation. Our manifest always wants to answer to our new passions, realizations, fulfilling real joys that arise within our new Eternal Sovereign Soul Presence Potential. The cosmos communicates through our new heart's excitement or passion sense about whatever we are willing to accept. Our imagination is always in its own delicious anticipation!'*

Indeed Bio-Masters, conscious evolution is free energy essence expression. And, **this free life force energy's creative fulfillment is the direct experience** of soul-spirit's organic qualities within the acceptance of ALL life. If you accept and are all life masters; then you can change anything, and miracles are as natural as sand!

LivingConcepts1.com

Essence Light Ambassadors Prepare the New Worlds 2020-2023- (predictions 11/2020 issue)
The New Ascended Masters- Maurene Watson

Masters, as changers of consciousness, your Master DNA communication codes are now activated to offer up your light Essence seed creations and visions to the world. This will now include *delicate bio-activations, to be the star-seed ambassadors you are, that interface with your new world star families and prepare the way for humanity. You will do this by living your creations and becoming them, thereby fulfilling your potentials.* This is transmitted to the world for others to receive these transmissions as potentials for their own unique awakening potentials as chosen by each soul. Hence, if another reads your book or takes your workshop or interacts with your energy field; their soul is activated and open to receive its next octave of light. *Your books, teachings, communications, and bio-codes are now living new matter.* They now exist as a living library of radiating plasma frequencies that act as a legacy like the Emerald tablets of Egypt, **(https://en.wikipedia.org/wiki/Emerald_Tablet),** that *held the secret of creating prima matt*er. The Telos-onium plates of ancient Lemuria which originated from Venus, served the same purpose. Each ancient cultural civilization in your Universe left their creations and records for humanity's ascension. Now you will do the same as a new species. It is your moment to pass the wand to humanity and move into your new passion potentials.

Since humanity has allowed itself to be deceived about the true cosmic history of its universe and who they are as a species; it's time to interface humanity with their divine memory's Essence energy communication networks they need, to find the brilliant solutions that only lighted love can offer. This is possible because your Light vessels have opened multi-quantum senses, multi-time, and multi-potentials in new free energy Essence DNA available to all. Instead of humanity fighting over issues that bind and separate, they will sense through your beautiful new Hearts potentials that offer new consciousness for *brilliant solutions in stewarding their New Earth colonies, as well as their own fulfillment.* As you open your new Essence vessels free energy sphere, you will transmit all the various ways that you have come into enlightenment which also reveals the truth of your cosmic history. Your new light-communication systems are a result of blending diverse cultures from all across the cosmos now being revealed in the new studies of: art, music, ritual, myth, science, technology, astrology, physics and medicine, to name a few. This comes with the realizations that these disciplines have been brought here by all the civilizations of your solar systems throughout your universes. It's time to integrate and upgrade the wisdom of the ages with your new visions and avail it for humanity's applications to heal themselves and their planet; *as well as be transmitted to all the new Earth colonies in their futures. This will be part of your new reality as you channel, interact, exchange, communicate, and attend council meetings in the New Earth realms and multi-light systems. You have long underestimated your agreement potentials with other universes and how you grow, learn, and are in constant exchange of Essence every moment. What your willing to accept and avail yourselves to the direct experience of as Human Masters, knowing your Essence can change or imprint any life form, is exchanged with all Life.*

These alternate versions of reality will provide the highest biological potentials available to every species on your planet. All these visions of how to change matter, and allow it to come alive as free energy are needed, as your world regenerates health and beauty through its own core bio-light vessel spheres. According to humanity's consensus reality, these multiple potentials were projected between 2038 through 2077; when conscious Divine-human essence Homosapien can teleport throughout the universe, create new matter directly from consciousness, and live hundreds of years in a natural bio-vessel. Can they accelerate potential outcomes due to your being the prototypes? This is where the **unknown gets to become manifest** because you are in a new cycle of evolution that will require bio-essence-enlightenment and/or bio ascension as the new standard. And, you are that new standard.

And as you know, each star seed must master their own soul template for this. And remember that Earth-Gaia is a Creator school for the new universes to both learn and inform the cosmos about soul Essence existence. So, as a part of this sovereignty, many of you are now and have been, being prepared in very *delicate bio-activations, to be the star-seed ambassadors to interface with your new world star families and prepare the way for humanity*. You are working with medical teams, bio-regenerative cosmic scientists, multi-light councils, and a bevy of cosmic beings to be prepared to carry these quantum frequencies in the Presence of your Being. These beings are also going through their own changes within the New Earth Super-Universe and learning from you as well.

Sovereign Bio-Master ascendants will channel their own composite integrated new Essence master code communications to ballast, trigger, and stabilize the migrations with all the new Cosmic Councils. Most Light Masters will channel an oversoul aspect of their new Cosmic group or their star families to fulfill their potentials and guide the youth. **Your youth**, with the average age on your world at 25 years; are already being activated to be the natural channels they are with the new DNA energy systems to complete their potentials. They will use technology as a tool till they create directly out of their own bio-consciousness. You have seen how easy it is for them to create once they release the ancestral DNA of their parents. They will continue to open their DNA codes and channels in ways the universes have never experienced in the new Divine Human imprint. And yet, there are many more already here, in every walk of life, that will reveal themselves and the systems they are form when appropriate that will amaze humanity with their stories. Indeed, you may interact **or channel inner contact in various ways**. Communication and guidance can come into your heart chamber channel via: video streams, dream realities, inspirations, invention ideas, traveling through your heart's star gate; direct manifestations; or through your beautiful new quantum essence trans-senses and new DNA code fractals. These star families can manifest into any form as required, just as you are being activated to remember and re-essence. Do not hesitate to ask other awakened channels and teachers on the planet to assist in your preparations as well as your own inner master teacher and new cosmic groups.

These communications will and have started in small groups and will grow across the globe till it is safe for your star families to interface with humanity directly. Until then, you will proceed, as you will be shielded from humanity's awareness by your high frequency energy spheres. You will **make the unseen worlds again seen** and it is part of yours and humanity's fulfillment. You have been channeling your most familiar families from Lyra-Vega, Orion, Pleiades, Andromeda, Sirius, Arcturus, etc. But many more systems and constellations will reveal now. Hence, your Creations, realizations, or chosen potentials in Divine Human embodiment on this New Earth, anchor and transmit options for all the emerging New Earth sovereign Creator Universes you Light Masters are creating on these New Earth Colonies. And many of you will serve as ambassadors and guardians as your channels are being prepared for such bio-diversity of messages, till humanity can accept their star families on mass. For now, you will safely re-essence and re-emerge your natural abilities and gifts that have seemed inaccessible.

This will **remind humanity's souls**, that if they remember their core Essence, then they can change themselves and their world with their own inner Essence consciousness. They can splice their own DNA without technology or ask the plant to grow them food that matches their cell essence? They will know humanity telepaths and holographs with relatives or star families who are no longer in human form, yet defy death; because everything is energy. Could they build homes out of living matter or materials that are alive and can adapt to changes in the weather? Could they re-engineer their land masses to support new bio-organisms that progenitor new life? What if their art sculpture was alive and could sing to them? What if there was a community cafeteria in every small town that raised its own food and where anyone could get a meal instead of having to buy dishes, shop for food, or drive to the market. What if they exchanged their soul gifts for currency? What if solar transponders powering from the moons, suns, and geothermal core were used to transfer energy instead of fossil fuels? How about bubble shoes made from recycled plastic made with an organic compound that fit the shape of the foot every time you wore them? They will come to know that the finite and infinite have remarried to produce new life that lives within their own consciousness. They will come to know that all technology and science can only mimic their essence consciousness. Indeed, outrageous change from multi-passionate free energy potentials that is in service to all life provides joyous living. And, you are that free energy life are you not? And yes, you will still love and help guide your universe without being pulled back, through bio-polarized emotional-mental reactions, or seductive illusions of the Old Earth matrix. *You are __beyond such vibrations__ now!*

In the free energy Essence bio-light vessel, conscious heart's awareness acts as an automatic compassionate action for all humanity. This is so, because love's communication is free energy to be used in any way the receiving soul's agency chooses to be activated by the bio-luminescence of your love's authentic Presence, in free Essence energy expression. Your illumination assists because it is outside all judgment, perception, and remains neutral and without harm or interference. Your living message remains that nothing exists, except in the consciousness of your own creation; and each is their own god-within no matter any external reality. Indeed, light's new Essence bio-vessel offers a return to natural organic standard of ethical consciousness. *Just by walking in your bio-luminescence, you transmit to* all those still suffering, dying, or caught in weather changes, the highest possible potential outcomes for all to choose from! You can now hold the quantum stable in your Essence vessels and begin your visions' fulfillments! There is no more waiting, your ascension is now to be lived fully in an unscripted life!

Humanity's Abandonment by the Gods Ascends- The New ascended Masters- through Maurene Watson 1/2020

Quantum Masters, your individual soul's Bio-ascension as your own unique creators is fully activated. As the pioneers of new Heart-DNA-essence consciousness, you are the template for the new quantum creation physics of new essence qualities of love within **ascended living.** All will experience cycles of transmutation symptoms on a daily basis that allow for immediate self-realized choices; as well as the vibrational stability of the Essence particle light core. Changing DNA cell consciousness of every life form in your world is the new norm, while harvesting new existences as Old Earth splits off into multiple realms and realities. Bio-ascension is authenticated by the direct free energy expression of unique, sovereign soul Essence DNA heart Presence, as the only Creator of its own embodied reality.

Humanity has been dealing with their extreme polarity creation memories and timeline existences while you were master coding the new DNA- essence heart imprint for the New Earth bio-light network systems. Eighty-percent of **Humanity has been lost** in, a collective unconscious/unawaken mind-emotion state of extreme anger, within projections of their: thoughts, feelings, attitudes, beliefs, and actions. *Their inner subconscious has felt abandoned, betrayed, rejected, and lost trust in their old dinosaur systems of: government, religion, money, and soul-less relationships and conditional loving and goodness.* They have been hypnotized to believe that the soul can be bought or sold by anyone or anything with or without responsibility or ramification of imbalanced energy exchange. Energy is free and must be balanced, and not trapped or enslaved. Humanity is beginning to release the old Earth DNA memory that their **existence is not angry abandoned punishment**, because of the consciousness you are transmitting. The underbelly of abandonment is unlimited potential of Life force energy. The life force energy is free energy, and is in an intimate relationship with each soul's inner Presence and the Eternal Presence of All life. Creative free energy is Life's Presence is in an energy relationship which is alive, passionate, expressive, sensual and full of intimate communication. Consciousness is sense-essence energy expression including all the qualities grown by the soul's DNA-heart. Free Energy lives in the Presence of all Life's existence. Everything is energy, and energy is everywhere. Hence, the new spirit heart is the **soul's DNA essence energy imprint,** igniting new code potentials for sovereign mastery.

However, **for humanity, the next decade** has come calling already. They will have to grapple with the choices of upgrading their bio-organic Essence human to enter the New Earths or accepting artificial chip implants, for Brain and body computer Interfaces. This may include connecting all humanity as one Tek-mind and as hybrids or Techno-humans; rather than activating their bio-conscious essence Homo-Sapien DNA-free energy vessel. Hence, time line memory warps and bio-particle accelerations are telling their soul-spirits that the vessel they occupy must have a base line DNA essence cell crystalline structure. This is so to make it safely through the New Earth star gates with a **light quotient infusion** of at least 52% stability. This includes continued acceleration, in order to naturally change essence for their next existence to continue on into their soul's bio-enlightenment, or move into the ascended realms. If not, the soul can then enter another evolutionary polarity cycle to continue finishing soul and oversoul contracts while serving the old energy universes for their bio-ascension. Indeed, the growth potential is enormous as it effects all humanoid, soul, and spirit families as well as their multi-verse outcomes. As the average age on your planet is 25, your next generations DNA is already activated to fulfill their DNA potentials for **mass migration not extinction.** Their Essence humanoid star families are waiting to interface with their Earth families for both learning and assistance. And, your humanity is full of extraordinary entities from all over the cosmos here in your Creator School Universe. There are master codes of every: DNA species, as well as color, sound, light frequencies that have ever existed making for extraordinary diversity within multiplicity. The hybrid ET-Universe and all its simulations will continue in parallel until these Creators and their manifestations have an opportunity to experience essence matter, its creation and its multi-free energy qualities in service to all life, rather than just forcing or trying to control evolution via simulated matter. Any life not en-souled will cease to exist

So, in these next five years, *humanity will discover the underbelly of their distorted/re-spliced DNAs' anger and rage that fuels their cycle of: fear, hurt, rejection, depression, loss of power, addicted mind; along with the shame, guilt, and grief of holding back their soul light. They must take back their lives, their choices, their biological responses, their true nature to love, and trust in the process of life again.* You are all too familiar with their creation myths, fallen god and angel stories, AI technologies, and hybrid aliens. For, you have ascended these through your new master codes into new DNA potentials. These old Earth Universe holograms may work for a while longer till Humanity discovers the true evolution of their species. However, what they will discover is equally rich.

They will finally feel or sense that they have been angry at everyone and everything, **except what they are really angry at. That is <u>primal DNA abandonment by Creation</u>** and their own spirit for seemingly allowing them to be traumatized or abused by other creators or aliens, as if they were 'thrown to the wolves'! This **<u>anger at Source</u>** is recorded in notorious stories/movies of separation between the spirit and: its soul, and its human; and by the loss of communication amnesia with the splitting or fragmentation of the creation families and their parallel aspects. It is recorded as fallen gods and angels who warred with ETs. However, **in full <u>potential this primal anger</u>** has transformed humanity into a new ascended species and taught compassionate love throughout the cosmos in ways never dreamed possible, even by the Ancients and Founders. So, as they release this ancient trauma and abuse; they are learning that a **healthy distance of boundaries from family and others is not to be confused with abandonment.** They will now remember to **<u>allow the abandonment its truth</u>** so their soul-spirit can re-sense, re-essence, or re-imprint angry abandonment at cell level. Then it can be changed into a new potential. *This is the right to your own soul sovereign: body, emotions, choices, responses, passion and* *the right to exist without have your empathic sensitivities violated or overburdened.* The Little human essence children of Earth did nothing wrong and made no mistakes. Now humanity can **stop abandoning their inner human child,** and remember their Divine Human communication inside the unconditional allowing of Father/Mother Creator parent within them, and their Creation. They can also allow their soul or spirit to embody in order to **re-essence nurturing and re-parent themselves and each other.** In other words, emotional empathic confusion and over identification with those who control their biological responses or sensitivities created a tyrant-victim slave experience. *This programing imprinted as an overwhelming distorted experience that felt like they had split off from their core essence threatening the existence of their very soul. This overshadowed their true essence knowing; that nothing exists or is real, except what is in one's essence soul consciousne*ss.

However, such polarized experience has offered the soul great evolvement via direct experience. Hence, you will see more courageous vast waves of light spreading in: towns, cities, nations, health arenas, and exchange commerce; all coming together to remind each other that they are not alone and can create a unified freedom that serves all life. <u>Unity consciousness ascends</u> through: Earth changes, death disease, poverty, suffering; by forced innovation, or exposures of systemic corruption and abuse within old Earth systems; and by the raising of the overall mass ratio of ascending light beings. This includes the children, and the ascendants all over your planet who have prepared the way by embodied living as Soul Sovereign free energy Bio-Masters; and transmitting that new consciousness across your worlds. Within this dialectic synthesis, Creation's Soul Essence Light Presence has begun to absorbs, filter, and ascend all humanity's shadows and distortions like a swirling magnetic rainbow. Indeed, now celebrate that humanity is ready to ascend primal abandonment into Essence nurturing of all life and watch its **incalculable outcomes these next years.** And this includes gradual interface with their star families who will or are already carrying an interface with the new DNA-Heart Essence Master codes.

Bio-Conscious Ascended Living Reviewed 2-2020 The New Ascended Masters- through Maurene Watson
Q: Can you please review how ascended living manifests?

Quantum Masters, we remind that your Bio-ascension is fully activated as the pioneers of new consciousness and the template for the new quantum creation physics of new essence qualities of love. All are experiencing cycles of transmutation symptoms on a daily basis that allow for immediate self-realized choices and changes. Changing cell consciousness of every life form in your world is the new norm, while harvesting new existences as Old Earth splits off into multiple realms and realities.

We also remind, that **Bio-ascension** is authenticated by the direct free energy expression of unique, sovereign soul Essence DNA Heart Presence, as the only Creator of its own embodied reality. Bio Ascendant awareness and self- realization replace: mind, emotions, gravity, time, polarity, power, energy and density. Embodied Self- realization replaces: perceptions thoughts, feelings, attitudes, and beliefs. Conscious self-realization and awareness replace these outdated learned and programmed systems. You are already everything that has ever existed even though you have allowed limited experiences of your own attributes and essence senses. Your human and divine senses have grown a new fabric of light in your new morphogenetic biosphere. Your *pioneering ascended heart* is transmitting into your world's awareness that you really live inside the interactions of your own consciousness without needing the *world's projections* of who you are. We also repeat, that your new cosmic star gates that align with the new Master DNA heart gate, have initiated quantum light that no organism has ever held inside bio-organic DNA-love before. And, each soul always progresses according to their soul's protocols and codes. *Are you ready to pioneer and live in your ascended pure energy Gem vessel, where you have birthed yourself as a new Cosmic Bio-Being in a new existence inside your very own consciousness?*

Let's review the **bio-ascension physics** for individual applications. **An Ascended Being** does not have polarity issue or wounds. An ascended being no longer uses their thoughts, feelings, attitudes or beliefs as medicine. They know that thoughts, feelings, attitudes and beliefs were all created from the biased intellect of judgment which replaced free energy experiences. Their conscious self- awareness is the medicine, sacrament, and ceremony for the world. They no longer need perceptions, ceremonies, mediators, or power objects or places to remember; or meditations to save the world. They know that their Soul's Core Heart Being is The Eternal Presence of everyone, everything/all life; and created though growing quality expressions of their sense Essences. They know their very existence is a living, loving sacrament to the world. They know the electro-magnetic chemical brain intelligence, data storage memory systems, endeavors of the mind or technology, are outdated systems. They are all replaced by new creations coded through the new DNA-Essence heart and growing quantum sense expressions of light, color, and sound. They manifest in new: technological applications, inter-relationship expressions, creative inventions, communication systems, multi-sense organisms, or super conscious creations for all life's eco-systems. They seeded these by mastering old Earth mind systems that regressed into polarity choices that were based on value judgments and biased comparisons from misappropriated energies of life forms unable to aware their own Creation codes. They also know that death, disease, and suffering are maladaptive over-learned experiences of human disconnected from its soul communication. There is **no separation** between their inner intimate communication with their inner soul-spirit Presence and the Presence of All Life! They no longer glamorize humanity's suffering or their wars as lessons that build soul character. They know they are simply choice distortions of *un-natural experiences* that were never resolved into freedom. The Ascended Being does not use challenges to rise above seeming limitations of other's realities. They do not live in any mind state of hypnotic acceptance where joy is sold as a commercialized product. They no longer try to perfect their human for its love has given their Divine Being rich experience and expression. They have absorbed their human wisdom back inside their Eternal Being. An ascendant no longer *projects* any reality outside their new torsion field biosphere. They have lived illusion and understand its limitations. Their DNA *cells* no longer register fear or limitation. They engage moment to moment in self- aware choices. ***They know, trans-essence-sense, aware, or intuit***; that distorted or projected outside realities feed as inflamed viruses created by dramatic stories that haven't yet returned to be loved by the Source that created them. All their experiential senses have authenticated that illusion is no longer acceptable in their creation. Allowing all life to be as it needs to be inside playful love, is like a theatrical art form for them.

All stages or cycles of the bio-ascent allowed transmutation of the flesh body and all physical reality matter density. During initial DNA reboots or re-splicing; it often sensed as trillions of inflamed alchemical elements were flushing the human's lymph's filtration system. It was the membrane dialysis of the sacred water molecule into: hot/cold>gases into liquid light> light into plasmas, >and quarks into>dark matter; which talk to the cosmos in particle light conversations. This allows the human, soul, and spirit to remember itself as One Being without any illusions or programming; and with the realization that separation was just an experiential time space distortion. Hence, an ascended being does not need to use power, control, energy, time, agenda, mind or mass to create because their Essence contains and IS these attributes already. Inside their cosmic egg or torsion sphere is heart's core consciousness that allows their Eternal Essence to create experiences of expression without end. They also know that their own bio-sphere can dissolve into free energy or pure essence in any moment such that an experience need never **be repeated, stored, or memorized, or re-incarnated.** The Master Human is in constant conversation with the cosmos and creation such that their consciousness can serve their free energy Essence expression moment to moment. They simply live in their Being-ness in a pure essence energy sphere using their trans-sense states (which replace thought and emotion) to accept raw experience. Light Master knows ***that their pure essence energy is the only authentic fulfillment of the unique soul's Presence inside the direct experience within their unique consciousness***. They know that there is nothing in this world or any other that they change, except their sense essence qualities of fulfillment within each moment of experience. These Beings know that they transmit these realizations that organic bio-cell essence can change anything as a sacrament to life. This is their gift of consciousness to the cosmos and presents an infinite unknown of potential outcomes on mass realities and all creations. They are aware, as a Free Energy Being; that their new plasma-particle bio-sphere stabilizes itself inside the heart DNA core. They live in a pure energy state; where all realms of creation inside their consciousness can be accessed. These quantum particles, transmitted from their heart chambers, in stable particle interactions create their reality moment to moment. Quantum interactions are always making love and creating from infinite unknowns, in an Eternal Self that breathes new expressive experiences.

In free energy Creation, in your new Master DNA codes, you are ascending into the free energy essence heart-DNA, and you understand that you're not your body. The vessel is a bio-imprint of the heart's DNA essence. Each unique Master's IAM soul, the IAMTHATIAM Spirit, and Sovereign Eternal Presence bio-Sphere, embodied as divine-human; is their Essence consciousness and the Essence expression of all life. The body shroud or animation is just a tiny light factor of spirit essence. You simply can't fit all your consciousness inside the vessel and that is why the master essence can change the Heart's DNA imprint into any life form chosen. This is why you see ascended masters using their robes as bio-spheres. This is because the DNA's heart's contains all the information of essence imprinting, not the brain or the body. They are just cells processors when programming is decoded or in detox, throughout the cell body. The essence DNA changes the body, not the sac of human electro-chemicals. The mind is a computer to process programmed memory systems for the human biology.

The progression to **grow new spirit essence** has been from: subconscious mind-emotions to>human feelings to>angelic senses> to Human-Divine spirit essence senses; growing into new multi-quantum qualities of rainbow essence spheres; or new qualities of love. The **soul essence** changes experiences moment to moment, once they are fulfilled. They are not programmed, recycled, memorized, recycled, or re-incarnated. Heart's self-awareness or self- realization can truly change your reality.

We repeat, that the Presence of Self Love, awareness, self-realization, or conscious potential; is created from a heart that can sense and communicate any experience inside itself and **essence immediate fulfillment**. And, this is an **intimately unique experience**, because you have co-created and are all life form*s*. Love's essence is a refined essence substance, within beautiful qualities of attribute expressions adding flavors to each new experience. This intimate communication with SELF in these intimate experiences with life grows the substance matter of consciousness via direct Essence sense experience. And, the substance matter or reality takes the form of life and that includes any blueprint for any life form. This ascended **inner heart channel** is able to communicate on all levels of existence at once to receive any expression that fulfills each soul in each Present moment. This is instant manifestation or direct matter manifestation right out of essence consciousness, as well as access to all other realms of existence.

As an ascendant, restoration of **Divine Memory** communicated to you that everything is this life has come to be about your enlightenment and Creator as a unique soul Sovereign Human Master. Each Master has their own bio-physics agenda for their natural progression for communication inside their bio-light network spheres, where the soul's tonal **DNA imprint** has fully activated itself in their new cycle of evolution. This organic imprint has been upgraded by the blend of human and divine senses merging into upgraded **super-sense intuition of heart's fulfillment**, and its bio-luminescent transmission to All Life. This is because enlightened conscious communication **bio-light systems** are available and functional for access. Their new bio-conscious DNA light network is the **new living code** for the intimate relationship between consciousness, energy, and creating new (dark) matter. Their bio-organism's DNA talks to the cosmos and the cosmos talks back in the form of instant communication and direct manifestation. Indeed, new multi or **trans-sense communication** with their inner Presence and all life, is the **master's code** for ALL life and the new ascending heart DNA transmissions.

These rainbow bio-spheres within spheres access new DNA codes, and their soul communication expressions can't be influenced by external people, places, circumstances, or events. This is an internal communion with all existence in every now. This new DNA heart-gate or bio conscious I-phone transmits the new standard of consciousness for all life cycles. This is a bio- light network communication inside the heart channel in/ on all realms of existence at once, without interference from any external reality whatsoever! Again, each moment inside their PRESENCE is fulfillment as its own instant manifestation. This makes the organic, Soul infused homo-sapiens' Essence, the most courageous and loving template to evolve in the cosmos thus far. This is conscious memory of the natural essence of bio-organic creation and that *Life always serves life no matter the experience*. Any perception is simply still a part of separation or distortion of experience; when it is understood that everything is pure consciousness.

Hence**, free energy** lives in the Presence of all Life. Life's Presence is in an energy relationship which is alive, passionate, expressive, sensual and full of intimate communication. Consciousness is sense-essence energy expression including all the qualities grown by the soul's heart love. Free Energy lives in the Presence of all Life's existence. Everything is energy, and energy is everywhere. Life is free energy or *passion in motion*. The new spirit heart is the **soul's energy imprint.** All Life animates its Presence in the Essence expression of free energy. Life is free energy communication and energy always seeks balance. You vibrate matter up or down or at multiples of the speed of light, color, and sound. The new DNA Heart Presence Essence can **change anything** via the excitation/passion of particle light waves. Conscious evolution in free energy qualities of love produces new super-intuitive senses of life and constant new potentials. The new DNA Heart Presence Essence can **change anything** via the excitation/passion of particle light waves. The sum total of your Essence-DNA light quotient determines access to your morphogenetic field. If the bio-body is not treated as sacred and lives in the programming of negative thoughts, feelings, attitudes, or beliefs of being: sick, contaminated, cell toxic, or locked in an opioid drug brain; **this limits access to higher frequencies** of your own morphogenetic energy field, or creating and merging with new rainbow quantum energy fields. Therefore, conscious evolution in free energy qualities of love produces new super-intuitive senses of life and **constant new potentials.** The **cosmos communicates through** new heart's excitement or passion sense about whatever they are willing to accept, change, image, or re-imprint. Hence, the bio-organism DNA genome is the anti-dote and immunity for all: viral, cyber, or DNA pathogens; and is now able to absorb cosmic radiation and adjust any time slips or space warps with your own consciousness. Indeed, they have restored any DNA bio-species distortions. Heart vibrates information of consciousness inside the new network light codes in your new DNA that excite your new qualities of love. *Eternal love's qualities of joy, new passions, renewed playfulness, fulfillment, transparency offer new octaves of conscious vibration.* There is always gravity in the heart's magnet to guide what matter imprints in the gravity of each situation. Included are natural potentials such as being one's own energy source. These masters can change matter, cell fusion and regeneration, tele-transmissions. They can you imprint and vibrate potentials in and out of different matter states of reality now; so that what matters to you is its own occurrence? They can travel on sound without need for a jet or beam from a spaceship. They can talk directly to souls or spirit beings without need for an I-phone as simulated consciousness? Their vessel is its own conscious technology and can imprint as any life form. So, in ascendant awareness, they remember and knowingly sense, that if they change the slightest choice in any moment, **all other potentials change**; always seeking the **most fulfilling choice.** Indeed, Essence expressions evolving nature is to: accept, allow, change, re-imprint, regenerate, fuse, re-essence, or re-molecule any experience. This is done through the new **DNA Heart Essence** imprint. *Now you know why we have called the heart chamber a: DNA- Essence Potential Source code/r, a bio-organic essence magnet, an atomic fusion accelerator, a stem cell replicator and regenerator, a quark centrifuge, a morphogenetic torsion field, and a midwife of worlds.*

So, the ascendant master code DNA-heart replicates any life form via a heart pulse. This happened because they transfigured the old limited human senses/pain body into new qualities of divine essence. For example, blue frequency could be: depression, a crayon, the sky, the soul, an aspiration, an invention: and yet, mixed with peach, it becomes a jellyfish that seeds new life in the ocean. **In the creative process,** the bio-organism simply follows its own codes and changes matter accordingly and this matter manifests in ongoing potentials as chosen reality. As evolution changes, that also informs and changes the cosmos. The master soul can dissolve any matter or substantiation of reality as no longer true, if it has been fulfilled, and energy no longer maintains or follows that focus. Or, it can remove Love's perception and its complete focus from programmed, patterned, or fulfilled realities that the soul has outgrown. Then, these perceptions or subtle judgments no longer exist and disappear, or dissolve inside one's conscious awareness back into pure essence. Then the soul returns to the still Eternal Potential Presence of the heart chamber, where new communication in a new moment awaits! The master soul always hears the **soul song** of their spirit heart saying, *'I love the way we trans-sense, and are intimate with the images, potentials, visions, expressions of our next instant manifestation. Our manifest always answers to our new passions, realizations, fulfilling real joys that arise within new Eternal Sovereign Soul Presence Potentials. The cosmos communicates through new heart's excitement or passion sense about whatever we are willing to accept'*

In sum, Bio- enlightenment is the resurrected birth of Heart's core light essence matter. Bio-ascension is the birth and ascent of new conscious quantum/prima matter (DNA codes) whereby the joy is the creation of bringing new essence potentials into existence. This is where soul evolution meets itself inside the consciousness of its own Sovereign Creation. Where the soul's evolution meets its own creation is the experience of fulfillment, because it is a new existence-experience from the soul's essence that is lived inside the Heart's core consciousness.

An Ascendant knows that all is well in the chronicles of the cosmos and it shares all its experiences with its cosmic star families. An ultra-terrestrial Human Master knows that an alien-**ET's greatest goal** is to become a conscious Bio-Essence Human. It knows that its pure organic Essence DNA expression has allowed all the *alien races* to heal and progenitor their own hybrid species. This allows the potentials for all the Old Earth Hybrid Universe to resolve all multi-cultural DNA's across the universes and heals all pasts and futures. It will lend understanding of **the organic essence human's ability to love**, to future DNA potentials for bio ascension, and the carriage of free will choices for their races. The Old Earth hybrid universe is still learning about the possibilities of growing love in differing matter states of bio-consciousness.

Final Reality Key for Ascendants 3 -2020
The New Ascended Masters- through Maurene Watson
Q: How do I embody or hold my Eternal Presence?

Bio-Light Masters, there remains only one reality key to live, embody, and self- realize now in your new life's unwritten script as you give humanity their Divine memory back. This is that **your Eternal Being** has now fully descended itself into a unique embodied creator. Descension and ascension have merged creating new bio-spheres of light. An ascendant sovereign Eternal being fully embodies the knowing; that they live, breath, and create, only, inside their own free energy essence consciousness, or biosphere. They remember and know through direct existence, expression, and animation as a spirit, soul, and human; that their very essence can change or create any life form, experience or creation inside their own organic biosphere. So, is there anything, absolutely anything, your allowing in your creation that you no longer need or you passion and is not there? Is there any place where you still externalize your creations? Yes, everything comes and goes from consciousness. Your consciousness is your unique soul-spirit's morphogenetic energy sphere in free energy expression and intimate expressive communication with all life. It is your version of creation and the All That Is. It is your: super universe, cosmic beings, ascended realms, masters, star beings, all constellations, universes; your New Earth, your unique soul-spirit-sovereign Eternal IAM Presence and the Presence of all life, which lives inside your own biosphere. *Everything that happens to you; you create, change, re-essence, re-imprint, or accept inside your own creational reality. Nothing exists outside of you, nothing.*

Your inner reality dictates and transmits any mirrored perceptions, reflections, self-realizations, or essence awareness; that are alive in your creation first. It does not matter what is going on in others creations including your partners, mothers, colleagues, friends, families, social media, planetary or even cosmic. Your responsibility is *to express and fulfill every possible potential* within your own creations via new direct experiences. *All parallel universes and realities can coalesce into one moment.*

This gets transmitted to all other creations and opens new corridors of choice for all creation. This is a quantum boon, *since everything in your world is being magnetized and absorbed into the light like a suction cup or quantum vacuum chamber.* If polarized matter is not freed, then it will cease to exist.

As graduates of creator school, you now live in new conscious communication bio-light network spheres within new qualities of infinite potentials to explore. You've self-realized that the divine memory of your soul heart's, new trans-dimensional Heart-DNA imprint, has fully trans-sensed its emergence in your new cycle of evolution. We reiterate again and again, that your new bio-organic DNA imprint has been upgraded by the blend of your human and divine senses merging into an upgraded super-sense intuition of heart's fulfillment, and its bio-luminescent transmission to All Life. This is because free energy Essence lives in the Presence of all Life. Life's Presence is in an energy relationship which is alive, passionate, expressive, sensual and full of intimate communication. Consciousness is sense-essence energy expression including all the qualities grown by the soul's heart love through both human emotions and divine senses merging. **This is the way of the passion of the heart.** Creation naturally fulfills all potentials of its very essence expression in evolving eternal realizations. As a new bio-genome life form or Eternal Being, passing though the Infinity gates into the New Earth realms, you are *growing endless qualities of love into infinite senses.* Divine memory lives inside the freedom of self-realized love and is only truly transformative through the direct experience of each unique soul. At Life's core, even the most heart wrenching experience is joy, because existence takes joy in just being alive and the divine right to express anything. This means that forgiveness no longer justifies love in the mind, but remembers and opens a direct path through the heart and can re-imprint new quality expressions and creations of love. This is the way of the heart.

Hence, you are Living Sovereign quantum masters now, who are mapping and embodying, your **plasma-particle-liquid Essence biosphere.** This new **Eternal Presence** Essence heart core light must be stabilized as you move into your new worlds.

This consciousness activates all light fusion systems till they are on auto pilot by mastering allowing natural organic life to self-regulate and self-generate evolution. Within this is the ability to hold the new core light stable in **quantum particle interaction**. During atomic-quantum interactions it seems that the cosmic egg's biosphere will explode, dissolve, or disappear as it oscillates between constant expansion and contraction before it reaches its **true stasis reality moment**. These masters often report that they are in and out of continual morphing. This is the quantum interaction between atoms and quarks creating new matter. These Quantum Masters essence themselves, like an undulating fluxing quark, in constant awareness of all potential choices at once. These quantum interactions continue, until their inner core's new star-sun plasma turns into particle light, and they live in the stability of their own **new conscious Biosphere**; without having to accommodate any holographic density **or geometric body. Hence, a stable motion of the free energy quantum master changing particle light, creates reality.** This is super-nova universal conscious access via the constant radiance of the individual core light of their Heart's sovereign new cosmic star-sun. As your Planet is doing the same, their very lives on this plane offer Eternity a new reality. For, Eternity has tasted the wines of freedom, choice, and embodied space-time existence. *This biosphere is the new free energy vessel or as the Particle Liquid Light or Quantum Matter bio-vessel.* Yes, you can practice allowing heart's awareness and love's reciprocity as you self- realize that all energy is free. You can allow changing potentials and realties manifesting in your now, as you live to the full potential of your core light, despite the periodic symptomatic morphing and energy undulations shifting. Your New heart Essence bio-vessel, which is trans-dimensional and trans-sensual in quality, has taken the rust off Divine cosmic memory. You remember now as cosmic adults who potential quantum consciousness inside new unknowns; that you can create, un/create, dissolve and re-essence any reality by choice. This will continue and change all perceptions your humanity holds of your worlds to transmit change.

In imprinting new consciousness in your bio-matter super universe, what is it like to walk in a vessel fabric of liquid plasma light matter? What does **particle light fusion** sound like (clairaudience/tele-sound), look like (clairvoyance/tele-video stream); touch, smell, taste, or sense soma like: (trans-sense / clairsentience), travel like, (tele-transport), or tele-com like? All occurs inside your own consciousness, where nothing happens unless you choose it and where each of you is your own creator-god. Essence light as your bio-fabric and multi-senses, replaces any old human: mental, emotional, physical or spiritual addictions or obsessions of thoughts, feelings or beliefs.

What is it like to **imprint your consciousness** on an object, an idea, a passion, a cell, or a new experience? What happens when quantum particles disappear and reappear? What happens when matter can change its own essence through freeing itself so that it might interact with life in any way it chooses? This new inner contact allows for a constant dialogue and conversation with the cosmos in all the spheres of quantum light. This **is not soul extension** in which the embodiment dies in an unconscious state or is locked in a nonphysical existence. Rather, this is **a soul-infused essence** embodiment in conscious, fluid transcendent states. Your biosphere replaces any holography or data patterns stored in the old earth DNA. Bio-essence goes into cosmic fusion where light sensors transmit and replace information once stored in the crown for open source access to new information in the moment it is needed via direct experience. These new imprints form fractal-DNA-unique source coding present in motion conscious movies, images, or trans-senses and are released into free energy the moment the new potential is fulfilled. As you build new matter in multiple potentials, your sensors interpret their essence in your light field vision as liquid plasma spiraling particles, floating on scalar waves along with all the trans-halo senses, colors, and sounds, so each experience is unique and free to imprint into any form or essence quality. We have coined these as new qualities of essence love that live within the Eternal Presence of The All That Is. Dark matter, or sovereign fusion, allows you to experience particle-light realities at will while still observing and interacting within this light universe without interfering because you only live in the composite bio-sphere of your own consciousness.

Therefore, the overall multi-**universal light resolution** with the interaction of the Old-Earth Universal story might simply be explained this way. When you first created this universe, your universe had its own cosmic-source sun Homo-Sapien DNA code organic physics. When the space-time warp ruptured through forced evolution and the past-present-future separated; twenty-three male and 23 female source sun creations collapsed. These source suns were to remain in their own creation blueprint codes. These creations Essence energy sphere physics and DNA-imprint codes were forced outside their natural flow. Creations outside your universe didn't all have organic/ magnetic essence bio systems to match. They had nuclear, electrical, and silicon mixtures of bio matter unsuitable to this universe that was trying to evolve itself into Essence. Their creations also fell inside and corrupted your universe. This **distorted the DNA** and each creation ended up trying to hybrid-ize enough essence via genetic manipulation, forced breeding, or arranged marriages to try to make their bio systems match. This was perceived as the **first acts of terror** or war due to misunderstanding of free will energy. This was an attempt to correct, control, or force, depending on the reaction of each Source Creation. It also created life extension or immortality which created death to the bio system instead of core essence regeneration, As a result, you still managed, due to your free will as an organic star seed with the original core essence Source coding blueprints for this Universe; to integrate and embody a new multi-DNA species bio consciousness by making a composite of all the other creations. You did this, **using Earth as a genetic creator school**. helping to heal, blend and integrate the other universes DNA codes within your universe despite their extreme reactions and diverse genetics. All the universes that collapsed can now heal their proper space time continuums and rebirth, regenerate, or dissolve by remembering their reactive stories. These universes have learned about the misapplication of freewill energies through force, power, control, judgment and interference with other source-code expressions; and are applying this wisdom via the New earth Super Universe you have all birthed. Because of the interbreeding and blending of creations, you have blended in their genetic source codes in this new organic/ magnetic heart stem cell, which all soul essence light beings will now have access, to ignite their core soul lights within. Each being has choices, depending on the light to mass ratio core soul essence light and its degree of evolution mastered during their existences. Now that this is a conscious light universe, your own inner Essence Source Creation Sun no longer has a shadow. Each master removed their shadow experiences as they finish old stories of limitation, judgment, time, polarity, and separation from Cosmic Self. Each light being ignites their own new master bio-matter codes for halo imprints as ready.

Gaia, along with her humanity, has birthed and imprinted billions of New Earths made of conscious dark or new essence matter. The New Earths already exist inside the light multi-verse of biospheres within biospheres. Humanity will continue to graduate out of the old cycle of evolution within the multi-verse ascension protocols. But they will use the old story experiences as wisdom to generate easier and more harmonious ways to evolve. In your biological enlightenment and going beyond physical reality in re-imprinting love's essence, you have become Quantum Essence Masters; where direct core experience is its own free energy core creation, its own intimate trans-essence expression, inside its own discovery of fulfillment. Each Essence is sovereign in free will and its right to evolve as its own Source with instinctual care for its own creations. Your bio-spheres of light now illuminate and transmit through the liquid light energy or quantum plasma particles. Your living Divine-Essence Human hologram transformed from liquid crystal to diamond-plasma and into quantum waves of particle light. Your cosmic egg has melted back into new dark matter Essence or light water bio-sphere. Your new bio-sphere Essence gives access to worlds, creative realms, and experiences you **have never lived before inside your own consciousness**. This ascendant essence Master Human knows its imagination is real. Its Essence Heart is real. Its Love is real and authenticated by direct expressive and intimate communication with in its own Eternal Presence of all life. Ascendant Human Master no longer works at being. Your bio-consciousness automatically creates itself through the excitement of its own imagination with immediate fulfillment. It sources, accepts, and allows Being at One with its own creative love, guiding the continued artistry of its own unique soul's expression. After all, it knows it is made of the essence matter of its own soul's existence and everything in life serves its conscious animation and excited exploration of creativity, fulfillment, expression, and self- discovery. There is always the natural pristine innocence of the unique Self to explore more of Essence Self inside creative love. **Creating your own reality** means full bio-integration to go beyond death and beyond physical human reality. Such is natural grace in existence, as your Human and its core light or 'soul', were never permanently designed to separate off from 'Itself,' fragment its bodies of experience, lose awareness of Itself; or be trapped on one plant such as Earth. Then the Q master has no inner or outer story that still lives within Old Earth hologram and enters the New Earth biophysical realities, and all else is released to the libraries of living antiquity. Here, the Essence Creator in you has adapted its animated new matter, such that, never again does it have to distort, mask, mediate, negotiate, compromise, or veil its experience. True, raw, authenticated experience is its own fulfillment. There is no needed meaning or purpose or justification for existence other than raw essence experience offered by cosmic evolution. Experience without a programmed anti-Essence is set free to be re-imprinted or dissolved.

This Consciousness cannot be fooled by the new **transhumance chip** that is being offered to humanity by your Old Earth controllers. Humanity is outgrowing the Old Earth matrix mind chip of historical beliefs, stories and perceptions that have enslaved them. Only continued terrorism to create chaos and the transhumant brain chip remain to keep the illusion of the matrix and make a cyber attempt to stop humanity's enlightenment. Only consciousness can re-imprint or re-essence life. However, brain chip technology claims to stem cell or DNA splice any disease, addiction, or human imperfection by logarithm codes and their replications. Conscious or Soul Essence codes can't follow DNA log rhythms' that are not organically essence unique to a life form's consciousness. Consciousness creates science and technology. *Technology-science cannot create essence consciousness, but it can be used as a benevolent tool to assist humanity as a tool till they awaken and grow their consciousness.* Technology can only translate or mimic awareness or provide machine learning, not re-essence life. Here there is not full consciousness; only a hybrid or replicated mix of organic essence with inorganic base atomic or quantum elements and particles. Technology can only offer the body healing or life extension, **not the soul**. However, the *new species sovereign Essence/soul codes are encrypted by each essence life form and require no chip*. Also, there is a danger, that a cyber human chip would short out or implode the brain or biological circuitry. So, the overall result of this enhanced human attempt can produce a hybrid, a clone, replicator, or robotic mixed with organic and synthetic compounds to mimic life. Conscious embodied Essence or Divine Humans are the only beings that can create any life form through the new Homo-sapien DNA species heart which imprints, or stem cells, through Essence love consciousness. Your new species of Q Essence enlightened humans will not barter their consciousness for technology? They remember being Galactic techno slaves once again through unfinished time lines or repeated solar wars and repeating that hologram in Atlantis. Against all odds, they have maintained their organic Heart's instinctual-ancestral Essence code roots in sovereign biological freedom?

However, <u>will awakening humanity</u> be intimidated or impressed by their governments fleet of spacecrafts that they reverse engineered from lower dimensional hybrid ETs; or turn to their own conscious light bio-physicals to create new worlds? There will be much more ethical or humanist BIOS, (bio-ecosystems input/output data), confusion and concern across disciplines, on what your multi-DNA-diverse species will become. Will organic Essence Divine Humans become the norm or will humanity choose some form of cyber transhuman? Will the quanta future of: Stereoscopic, augmented, or halo sensor realities, become robotic or human machine mimics of the healing timelines of Atlantis? Or will the Quantum Conscious Essence master, **living within its own unique Eternal Presence inside its sovereign bio-sphere and fed by quantum light, become** the new standard? Here, the heart's imagination, not the distorted mind, can create any reality that outdates technology?

In the final reality key for the ascendant, this bio-Essence-integration completes as the New DNA-Species Heart expresses and fulfils in new unknown potentials. That includes going into Divine Human beyond physical reality to create your own Conscious Essence realms and realities. And, you do this as your own Eternal presence; or as Source Core Creator in free energy dark matter Sovereign Spherical Consciousness. All that is not you, all that is not love, and all life that is held in separation from creation, dissolves and ceases to exist. Quantum Core Essence as your Eternal Sovereign Presence creates out of its very own consciousness any life form into new realities or experiences; never embodied by a full Essence Human prior in the cosmos. The cosmos channels your love into **Unknown new adventures** that you can choose to share with others and the rest of the cosmos! In your new meta universes there will be infinite ways to travel, create, express love, and find fulfillment evolving from your continued Quantum progression in your embodied Eternal Presence. Again, Enlightenment requires you lose/reprogram your mind-emotions to go beyond them. Enlightenment is impossible through the mind. It comes through the Essence DNA-heart. Hence, no one leaves you in depression or loss, and you don't leave anyone in anger or hurt. Each of you **just owns more of yourselves** from the reflection of shared experience. Sovereign is love's essence and that is why you never lose anyone or anything. You just get a deeper sense essence expression of your Presence in reflection and absorption of more of what you already are. But that sweet, tasty sense of intimate, fulfilled, intimate communication with life's actors shared between souls; imprints all life with one impeccable, indelible glance as a new potential. And, in your enlightenments you've had to: channel, write, teach, and re-grow your souls out of the blind crazy mind to sing, dance, and play your own new soul essence song again!

How again did **judgment potential justice?** Human's new bio-DNA species essence compassion has grown the old Earth pain body of fear, loss, guilt, and limitation into new essence-sense qualities of joy, passion, fulfillment, transparency and inner communication. All trapped or extreme energies can be **released into new potentials**, because the new bio-essence DNA-codes now operates within a free energy light communication system of your own Master Presence within all life. Militaries and governments can open free energy technologies for peace. Cultured animal, plant, and crystal essence cells can be **incubated or re-imprinted** for new food sources without killing life force. The master code stem cell re-splices/regenerates the essence DNA to heal any disease or re-grow an organ. Organic Cyber sensor-chips can answer to the bio-organics consciousness of the human species, to avert disasters, cyber entrapment, or genocide. So, what is this free energy essence expression that communicates in service to all life? Free energy lives in the Essence Presence of all Life. Life's Presence is in an energy relationship which is alive, passionate, expressive, sensual and full of intimate essence communication within the natural essence of each unique soul.

You have learned to create/architect, destroy, and rebuild a physical free will- marriage-family, physical polarity space- time universe into a new multi-quantum bio-enlightened universe. You have even pulled all your old energy universes with their potentials and parallel realities into One new multi-diverse Creator reality, by becoming your own ascendant unique Conscious Creator. New Earth School allows you the final reality keys for ascension. That is, to create from inside your own unique embodied Eternal-Essence Presence, new bio-imprints, worlds, realms, universes, and everything else your consciousness can imagine!

Master Aspect Potential Ascends 4-2020

There seems to still be some confusion about the **composite Integrated Divine Self** that has collected all its aspects into your new Divine Human Master Creator. Bio-Light Masters, there is a major difference between old limited energy and new energy creational aspects including: lifetimes, life forms, bodies, patterns, blueprints, emotions, existences, or even senses you have lived. You either transform the Old Earth universal aspects or set them free while integrating their wisdom by reviewing their movies. This is what enlightenment or self- realization offers you. The Ascendant Master Creational Self then uses their wisdom into new energy creations that are playful and fulfilling trans-sensual expressions of any potential imaginable; for that is what evolutionary Creators love to do!

As you well know and have lived; these limited old energy polarized aspects lived in patterns, DNA splices, wounded mind-emotions and bodies of separation. In separated states they experienced death, disease, suffering, and every: shame, blame, doubt, judgment, regret, and anti-love/life, or alien life form possible. This was to grow the soul's new DNA biosphere into a sovereign anti-pathogenic, anti-cyber, anti-time warp, free energy agency. Indeed, it was so that the soul could evolve its own unique love. Don't forget that each limited experience of these human and soul aspects who lived unspeakable acts and unnatural practices are the ones who **grew your new essence potentials**, for your new DNA- Heart. They are loved for their service because Creation loves all its creational aspects and evolves from their storied experiences. For, all these old energy aspects have grown new qualities of Essence love.

So, part of enlightenment, is **the constant self-realization** from the new spirit now animating the vessel communicating through your composite Integrated spirit. Your master Spirit is telling you, *"thankyou"* for embodying me as a: plant, animal, male/female human, child, the wind, a stone, a reptile, an angel, a blueprint pattern, an energy essence, and creator. Such is the process of master imprinting. If you have been a physician, then Integrative Spirit Self can heal your cold. If you were a space commander, then you remember your brothers and sisters in the stars. In your divine memory as an angel, you know you can re-imprint any life form with your essence. If you were in Atlantis, then technology is an easy tool for you now. If you studied in the mystery schools then all the ancient wisdom is available as to why you are here on Earth now. If you destroyed planets, then you understand the physics of atoms in creation, destruction, and regeneration. If you were a midwife crystal maiden; you birthed new DNA children that would ensure natural evolutionary life cycles. If you were an Elohim, then you know that you have charge over the elemental imprints of the forces of nature and the power of the elements. And when you remember that you have always been a spirit being, then access to the higher realms is always available

Hence. the import is that this composite self-realized or True Self aspect is constantly drawing on the wisdom of all your experiences to draw out new essence potentials in a continual integrative dialogue. This new dialogue is the voice of your Holy Spirit Presence guidance consciousness in self-care, self-love, or passion playing with its fulfilling expressions/essence potentials; and grounding in your new free energy vessel. Because it is human and divine merged, it keeps you from recycling a pain body past or future you have already lived. It also enjoys chocolate cake, human skin, nature, relationships and all the intimate sensuality creative life has to offer moment to moment with a freelance energy script. Inclusive is full access to the quantum realities and realms. This True Self now holds or embodies you in your new vessel long enough to complete all possible potentials in the New Earth realms. It also changes the channel when you are no longer the star of your own movie. Presence will also continue to guide you deeper and deeper into your own consciousness until you are fulfilled to the brim with every potential you can imagine playing with. Indeed, your new innocence gets to play with all the new toys in creation. Indeed, its internal voice of communication awareness is instrumental in reminding you that you are now here to enjoy the love you are in all new realms of life. So, no matter what is going on in the worlds around you, no matter what goes on with your family, relationships or friends, and no matter what you see outside your own bio-sphere; your sovereign master creator has guaranteed that you will be fulfilled if you allow its full embodied Presence to play with your new manifesting love. We remind that this is moment to moment in your creational process that changes you and your cosmos. Again, the cosmos hears your *elegant voices* of inner master code communication from your Spirit Heart. *"I love the way we can live together in our moment in that empty space in our heart. I love the way we trans-sense, and are intimate with the images, potentials, visions, expressions of our next instant manifestation. Our manifest always answers to our new passions, realizations, fulfilling real joys that arise within our new Eternal Sovereign Soul Presence Potential. The cosmos communicates through your new heart's excitement or passion sense about whatever you are willing to accept, change, image, or re-imprint. Heart vibrates information of consciousness inside our new network light codes in our new DNA that excite our new qualities of love. All you have to do is sit together in the heart chamber in our new Divine Bio-Human vessel and the cosmos will transmit the image, new quality sense or re-imprint our DNA into whatever experience you want to become."*

www.ingramcontent.com/pod-product-compliance
Lightning Source LLC
Chambersburg PA
CBHW030931180526
45163CB00002B/535